Felix Heckendorn

Control of Sheep Parasites

Felix Heckendorn

Control of Sheep Parasites

The Control of Gastrointestinal Sheep Nematodes with Tanniferous Forage Plants

Südwestdeutscher Verlag für Hochschulschriften

Impressum/Imprint (nur für Deutschland/ only for Germany)
Bibliografische Information der Deutschen Nationalbibliothek: Die Deutsche Nationalbibliothek verzeichnet diese Publikation in der Deutschen Nationalbibliografie; detaillierte bibliografische Daten sind im Internet über http://dnb.d-nb.de abrufbar.
Alle in diesem Buch genannten Marken und Produktnamen unterliegen warenzeichen-, marken- oder patentrechtlichem Schutz bzw. sind Warenzeichen oder eingetragene Warenzeichen der jeweiligen Inhaber. Die Wiedergabe von Marken, Produktnamen, Gebrauchsnamen, Handelsnamen, Warenbezeichnungen u.s.w. in diesem Werk berechtigt auch ohne besondere Kennzeichnung nicht zu der Annahme, dass solche Namen im Sinne der Warenzeichen- und Markenschutzgesetzgebung als frei zu betrachten wären und daher von jedermann benutzt werden dürften.

Verlag: Südwestdeutscher Verlag für Hochschulschriften Aktiengesellschaft & Co. KG
Dudweiler Landstr. 99, 66123 Saarbrücken, Deutschland
Telefon +49 681 37 20 271-1, Telefax +49 681 37 20 271-0, Email: info@svh-verlag.de
Zugl.: Zürich, ETHZ, Dissertation, 2007

Herstellung in Deutschland:
Schaltungsdienst Lange o.H.G., Berlin
Books on Demand GmbH, Norderstedt
Reha GmbH, Saarbrücken
Amazon Distribution GmbH, Leipzig
ISBN: 978-3-8381-0133-0

Imprint (only for USA, GB)
Bibliographic information published by the Deutsche Nationalbibliothek: The Deutsche Nationalbibliothek lists this publication in the Deutsche Nationalbibliografie; detailed bibliographic data are available in the Internet at http://dnb.d-nb.de.
Any brand names and product names mentioned in this book are subject to trademark, brand or patent protection and are trademarks or registered trademarks of their respective holders. The use of brand names, product names, common names, trade names, product descriptions etc. even without a particular marking in this works is in no way to be construed to mean that such names may be regarded as unrestricted in respect of trademark and brand protection legislation and could thus be used by anyone.

Publisher:
Südwestdeutscher Verlag für Hochschulschriften Aktiengesellschaft & Co. KG
Dudweiler Landstr. 99, 66123 Saarbrücken, Germany
Phone +49 681 37 20 271-1, Fax +49 681 37 20 271-0, Email: info@svh-verlag.de

Copyright © 2009 by the author and Südwestdeutscher Verlag für Hochschulschriften Aktiengesellschaft & Co. KG and licensors
All rights reserved. Saarbrücken 2009

Printed in the U.S.A.
Printed in the U.K. by (see last page)
ISBN: 978-3-8381-0133-0

Acknowledgements

I am very grateful to PD Dr. med. vet. Hubertus Hertzberg for his continuous encouragement and confidence, which made the completion of this work possible. I highly appreciated his kind way of responding to the thousands of special questions I had during the last four years and I am also thankful for his support during the intense phases of writing articles and the thesis presented here.

I am indebted to Dr. Veronika Maurer for vigilantly leading the 'Veterinary Parasitology' research group at the Research Institute of Organic Agriculture (FiBL). She continuously prepared grounds for a very flexible and autonomous working atmosphere, which permitted the realization of the laborious feeding experiments required for this thesis. In this regard I also thank the other group members Erika Perler, Zivile Amsler and Ilse Krenmayer for their irreplaceable help in the field and in the lab.

Professor Dr. Wolfgang Langhans was responsible for the overall supervision of this thesis. I highly appreciated his very efficient and helpful input throughout my PhD period. Together with Dr. Markus Senn he guaranteed a perfect guidance and the meetings periodically held throughout this work were always fruitful and motivating. With respect to the papers and other scientific documents, the comments and suggestions of both Professor Dr. Wolfgang Langhans and Dr. Markus Senn are highly acknowledged.

This thesis represents one of three dissertations performed in the framework of an interdisciplinary Swiss project on tanniferous forage plants. The experiments presented here would not have been practicable without the help of Dieter Adrian Häring (PhD thesis at the Swiss Research Institute ART Reckenholz), who covered many aspects of tanniferous plant cultivation and Anna Scharenberg (PhD thesisat the Swiss Research Institute ALP Posieux), who took care of feed analysis and referenced in adequate animal feeding. In this context I would like to thank the whole steering group of the project: Professor Dr. Michael Kreuzer, Dr. Frigga Dohme, Dr. med. vet. Andreas Gutzwiller, Yves Arrigo, Dr. Andreas Lüscher, Dr. Daniel Sutter, Professor Dr. Niklaus Amrhein and PD Dr. Peter Edwards.

Professor Dr. Michael Kreuzer supervises the thesis underway in the area of animal nutrition and he kindly accepted to co-examine my thesis. His support and helpful suggestions are highly appreciated.

Several parts of this work where performed in close collaboration with the Institute of Parasitology, Zurich (IPZ). I would like to express my gratitude to the head of the Institute, Professor Dr. med. vet. Peter Deplazes for having made this possible. Dr. Paul Torgerson and Lucia Kohler (both from IPZ) helped with statistics and in the lab, respectively.

Dr. Steve Wilcockson (University of Newcastle, UK) kindly improved the English of the publications and conference abstracts originated from this work.

I got to know many people in the last four years, indirectly or directly contributing to my thesis. The input of the following is particularly acknowledged: PD Dr. Jörg Heilmann, Dr. Hans Leuenberger, Dr. med. vet. Janina Demeler, Dr. Ronald Kaminsky, Peter Ackle, Pius Allemann (and his team), Samuel Schaffner (and Teddy), Kathrin Bühler, Philip Graf, Sabrina Schaller, Hans Ursprung and Jakob Zinsstag.

Finally, I would like to thank:
My girlfriend and best friend, Rezia Buchli for the support and love she gave me in the course of my PhD work.
My mother Jacqueline Grosskopf and my father, Werner Heckendorn for their generosity and care which made my education possible.
All my friends, who supported me through all up's and down's, for their friendship, love, support and understanding; namely: Basil Heckendorn, Palabro Seguridad, Urs Berner, Grischa Madörin, Philipp Hördegen ... and many others.

Table of Contents

1.	Summary	5
2.	Zusammenfassung	7
3.	Introduction	10
3.1.	Alternative approaches to control the free-living-stages of gastrointestinal nematodes	10
3.1.1.	Grazing management	10
3.1.2.	Biological control	11
3.2.	Alternative approaches to control gastrointestinal nematodes within the host	11
3.2.1.	Genetic resistance	11
3.2.2.	Vaccines against gastrointestinal nematodes	12
3.2.3.	Herbal anthelmintics	12
3.2.4.	Gastrointestinal nematodes and host nutrition	12
3.3.	Tannins	14
3.3.1.	Biochemical structures and properties of tannins	14
3.3.2.	Tannins in ruminant fodder plants	14
3.3.3.	Nutritional and animal production aspects of tanniferous plants in non-infected ruminants	15
3.3.4.	Antiparasitic and physiological effects of tanniferous plants in infected ruminants	15
3.3.5.	Mechanisms of condensed tannin action	16
3.3.5.1.	Indirect effects	16
3.3.5.2.	Direct effects	16
3.3.6.	Variability of anthelmintic effects of condensed tannins	20
4.	Goals of current research	22
5.	Individual administration of three tanniferous forage plants to lambs artificially infected with *Haemonchus contortus* and *Cooperia curticei*	24
5.1.	Introduction	24
5.2.	Animals, Materials and Methods	26
5.2.1.	Forage cultivation	26
5.2.2.	Animals	26
5.2.3.	Parasite isolates and experimental infection	26
5.2.4.	Experimental design	26
5.2.5.	Forage administration, feed intake and live weight	26
5.2.6.	Feed analysis	27
5.2.7.	Parasitological procedures and measures	27
5.2.8.	Statistical analysis	28
5.3.	Results	29
5.3.1.	Botanical feed analysis	29
5.3.2.	Physical and chemical feed analysis	29
5.3.3.	Live weight, feed intake and faecal output	30
5.3.4.	Faecal egg counts (FECDM) and daily egg output (TDFEO)	31
5.3.5.	Worm burden	32
5.4.	Discussion	33
5.4.1.	Are differences of condensed tannin action against gastrointestinal nematodes related to the host organ?	33
5.4.2.	Plant specific anthelmintic activity of condensed tannins	33

5.4.3.	Is there an in vivo dose-response relationship?	34
5.4.4.	Reversibility of parasitological effects	36
5.4.5.	Interpretation of faecal egg count can be ambiguous in feeding trials	36
5.5.	Conclusions	37
6.	Effect of sainfoin (*Onobrychis viciifolia*) silage and hay on established populations of *Haemonchus contortus* and *Cooperia curticei* in lambs	38
6.1.	Introduction	38
6.2.	Materials and Methods	38
6.2.1.	Animals	38
6.2.2.	Forage and feed constituents	39
6.2.3.	Parasite isolates	39
6.2.4.	Experimental design and measurements	39
6.2.5.	Faecal samples and culture processing	39
6.2.6.	Statistical analysis	40
6.3.	Results	40
6.3.1.	Nutritional contents and condensed tannin concentrations	40
6.3.2.	Consumption of feeds and live weight gain	40
6.3.3.	Faecal egg counts	41
6.3.4.	Worm burden and per capita fecundity	41
6.3.5.	Packed cell volume	41
6.4.	Discussion	45
7.	On farm administration of sainfoin (*Onobrychis viciifolia*) silage to ewes naturally infected with gastrointestinal nematodes: effect on periparturient egg rise	47
7.1.	Introduction	47
7.2.	Materials and Methods	47
7.2.1.	Animals	47
7.2.2.	Forages	47
7.2.3.	Experimental design and measurements	48
7.2.4.	Statistical analysis	48
7.3.	Results	48
7.3.1.	Consumption of feeds and condensed tannin concentrations	48
7.3.2.	Faecal egg count and 'area under the curve'	48
7.3.3.	Larval cultures	49
7.4.	Discussion	50
8.	Overall discussion and Conclusions	52
8.1.	Importance of condensed tannin dose	52
8.2.	In vivo dose-response relation	53
8.3.	Problems related to faecal egg counts	53
8.4.	Further research addressing the variability of anthelmintic responses	53
8.5.	Inclusion of tanniferous fodder plants in rational farming systems – proposed research	54
8.6.	Towards an integrative approach of alternative gastrointestinal nematode control	54
9.	References	55
10.	Publications	61

1. Summary

Infections with gastrointestinal nematodes (GIN) represent a major constraint in sheep husbandry. For many years, the control of these parasites has solely relied on the repeated use of synthetic anthelmintics. However, the emergence of resistant GIN populations and the increasing concern of consumers for drug residues in animal products have provided a strong impetus towards the development of alternative strategies to control GIN. Amongst those, the administration of tanniferous fodder plants to infected animals received prime attention in the last decade, and although a considerable body of knowledge on this strategy has been accumulated, many questions remain to be answered.

In sheep and goats, the anthelmintic effects observed in response to tanniferous fodder administration are believed to be associated to condensed tannins (CTs), contained in the plants. Chemically these molecules belong to the polyphenols and are expressed by a range of fodder plants together with other secondary metabolites. Across experiments, the most consistently observed anthelmintic effect is a reduction in parasite egg output (as measured by egg counts per gram faeces – FEC). Reductions in adult or juvenile GIN as well as decreases in female worm fecundity have also been reported but were less consistent, varying with the administered tanniferous plant species, the nematode species and also the host species (i.e. sheep or goat). The mechanism of CT-action on GIN is unknown, but it has been hypothesised that the high binding capacity of CT-molecules with protein might be involved in causing the effects. The investigation of the different sources of variability associated with the antiparasitic effects represents the major challenge in this area and some aspects of this research are covered in the present thesis.

In a first feeding experiment, we investigated direct anthelmintic effects associated with the feeding of fresh herbaceous tanniferous forages and chicory against established populations of *Haemonchus contortus* and *Cooperia curticei* in lambs. Twenty-four parasite naïve lambs were inoculated with a single dose of infective larvae of these parasites 27 days prior to the start of the feeding experiment. Lambs were individually fed with either chicory (*Cichorium intybus*), birdsfoot trefoil (*Lotus corniculatus*), sainfoin (*Onobrychis viciifolia*) or a ryegrass / lucerne mixture (control) for 17 days. Animals where then united to one flock and subjected to control feeding for another 11 days to test the sustainability of potentially lowered egg excretion generated by tanniferous forage feeding. When compared to the control feed, the administration of all tanniferous forages was associated with significant reductions of total daily faecal egg output specific to *H. contortus* (chicory: by 89 %; birdsfoot trefoil: by 63 %; sainfoin: by 63 %; all tests $P < 0.05$) and a tendency of reduced *H. contortus* worm burden (chicory: by 15 %; birdsfoot trefoil: by 49 % and sainfoin: by 35 %). Irrespective of the experimental fodder, no anthelmintic effects were found against *C. curticei*. Cessation of CT-feeding followed by non-CT control feeding did not result in a re-emergence of FEC based on faecal dry matter (FECDM) in any group, suggesting that egg output reductions are sustainable. The moderate to high concentrations of CT in birdsfoot trefoil (15.2 g CTs kg^{-1} DM) and sainfoin (26.1 g CTs kg^{-1} DM) were compatible with the hypothesis that the antiparasitic effect of these forages is caused by their content of condensed tannins. For chicory (3 g CTs kg^{-1} DM), however, other secondary metabolites need to be considered. Overall, birdsfoot trefoil and sainfoin seem promising candidates to contribute to the control of *H. contortus*, not only by mitigating parasite related health disturbances of the host but also by a sustained reduction of pasture contamination.

Experiment two was carried out using a similar set-up as in experiment one (i.e., established populations of *H. contortus* and *C. curticei*, 16 days of experimental feeding). The objective of this second study was to examine the anthelmintic effect of dried and ensiled sainfoin (*O. viciifolia*). Twenty-four parasite naïve lambs were inoculated with a single dose of infective larvae of the nematodes 28 days prior to the start of the feeding experiment. Groups A and B received sainfoin hay (61 g CTs kg^{-1} DM) or control hay (1 g CTs kg^{-1} DM) for 16 days. Groups C and D were fed on sainfoin silage (42 g CTs kg^{-1} DM) or control silage (1 g CTs kg^{-1} DM) for the same period. Feeds were offered ad libitum and, on the basis of daily refusals, were supplemented with concentrate in order to make them isonitrogenous and isoenergetic. The consumption of conserved sainfoin was associated with a reduction of adult *H. contortus* (by 47 % in the case of hay, P < 0.05; by 49 % in the case of silage, P = 0.075), but had little effect on adult *C. curticei*. Compared to the controls, *H. contortus* specific FECDM was reduced by 58 % (P < 0.01) in the sainfoin hay group and by 48 % (P = 0.075) in the sainfoin silage group. For both sainfoin feeds FECDM specific to *C. curticei* were significantly decreased when compared to the control feeds (hay by 81 % and silage by 74 %, both tests P < 0.001). The data of this study suggest that different mechanisms were responsible for the reduction in FECDM in response to feeding tanniferous fodder. For *H. contortus* the decrease seemed to be due to a nematocidal effect towards adult *H. contortus*. In contrast, for *C. curticei* the reduction in FECDM appeared to be a result of a reduced per capita fecundity. For both, hay and silage, an antiparasitic effect could be shown, offering promising perspectives for the use of conserved tanniferous fodder as a component of GIN-control.

The aim of the third study was to investigate the anthelmintic effect of sainfoin silage fed as a sole diet on a mixed GIN-population in periparturient ewes in an on-farm setting. Thirty-three ewes with FEC > 300 were divided into two groups of 16 (sainfoin sialge) and 17 (ryegrass / clover silage; control) animals and fed on the respective feeds for twenty-five days. The GIN-population predominantly consisted of *H. contortus* (~ 50 %) and *Trichostrongylus colubriformis* (~ 35 %). After ten days of consecutive sainfoin silage feeding, FECDM was reduced by 60 % (P < 0.01) when compared to the control fed animals. FECDM of the sainfoin fed group consistently remained lower until the end of the study compared to the control. The area under the curve (AUC) based on FECDM, calculated for the entire experimental period was lowered by 32 % in the sainfoin group but this reduction was not significant (P = 0.17). When the AUC was calculated for the period following FECDM reduction in the sainfoin group (i.e. day 10-24) only, the reduction was 65 % (P = 0.09). The results from this study confirm some favourable antiparasitic effects associated with the consumption of sainfoin silage in naturally infected periparturient ewes. The potential impact of this effect on the epidemiology of trichostrongylidosis needs to be critically evaluated.

In conclusion, our results suggest that the antiparasitic effects of tanniferous forages in general are achieved at lower CT-levels in the abomasum than in the small intestine and therefore would rather be organ dependent than GIN-species related. It has furthermore been shown that conservation of tanniferous fodder plants does not alleviate the anthelmintic effects and therefore offers exciting opportunities with respect to centralized production, sale, storage and an extended administration independent of the season. To date, however, the approach of tanniferous forage administration cannot be expected to provide satisfactory sustainable control of GIN. The combination with other alternative strategies in the sense of an integrated approach of GIN-control might be promising in terms of producing additive effects. Until rational integrative control approaches are available, the complete abandonment of synthetic anthelmintics is not sensible. However, in view of the widespread resistance of GIN against these drugs their economical and carefully targeted use is of vital importance until integrated alternative parasite control strategies are available.

2. Zusammenfassung

Der Befall mit Magen-Darm Nematoden (MDN) stellt beim Schaf einen wesentlichen limitierenden Faktor in der Haltung und Produktion dar. Während vielen Jahren stützte sich die Behandlung und Kontrolle dieser Parasiten auf chemisch-synthetische Entwurmungsmittel (Anthelminthika). Das gegenwärtig vermehrte Auftreten von anthelminthika-resistenten MDN Populationen sowie das verstärkte Bedürfnis der Konsumenten nach medikamentenrückstandsfreien tierischen Produkten haben der Forschung in diesem Bereich einen starken Anstoß zur Entwicklung alternativer Kontroll- und Behandlungsstrategien gegeben. Eine dieser Strategien stützt sich auf die Verfütterung tanninhaltiger Futterpflanzen zur Kontrolle von MDN. Diese Strategie gewann im letzten Jahrzehnt an Beachtung. Obwohl gegenwärtig eine beachtliche Menge an Forschungsresultaten zum Thema vorliegt, bestehen noch zahlreiche offene Fragen.

Aufgrund der bisherigen Erkenntnisse wird davon ausgegangen, dass die anthelminthische Wirkung tanninhaltiger Pflanzen auf die in ihnen enthaltenen kondensierten Tannine (kTs) zurückzuführen ist. Nebst anderen sekundären Metaboliten werden diese zu den Polyphenolen gehörenden Moleküle von verschiedenen Futterpflanzen exprimiert. Über alle verfügbaren Resultate betrachtet, ist die Reduktion des parasitären Ei Ausstoßes (gemessen in Anzahl MDN Eiern pro g Kot — EpG) der am konsistentesten beobachtete anthelminthische Effekt. Die in Folge der Verfütterung tanninhaltiger Pflanzen gemessenen Reduktionen der juvenilen und adulten MDN Stadien waren weniger konsistent und variierten mit der verabreichten tanninhaltigen Pflanzen-Spezies, der/n untersuchten MDN-Spezies und der in die jeweilige Untersuchung eingeschlossenen Wirtstierart (d.h. Schaf oder Ziege). Obwohl der Wirkungsmechanismus der kTs gegen MDN unbekannt ist, wird davon ausgegangen, dass die hohe Affinität der kTs zu Proteinen dabei eine wichtige Rolle spielt. Die Klärung der verschiedenen Faktoren die für die Variabilität der anthelminthischen Effekte verantwortlich sind stellt in diesem Forschungsbereich die größte Herausforderung dar. Die Vorliegende Dissertation nimmt sich einiger der aktuellen Fragen an.

In einem ersten Fütterungsexperiment wurden die mit der Verfütterung von frischen (d.h. unkonservierten) tanninhaltigen Futterpflanzen im Zusammenhang stehenden direkten anthelminthischen Effekte auf adulte *Haemonchus contortus* und *Cooperia curticei* untersucht. Siebenundzwanzig Tage vor Beginn des Experiments wurden 24 parasitennaive Lämmer einmalig mit infektiösen Larven dieser Parasiten inokuliert. Die Lämmer wurden während 17 Tagen mit entweder Chicorée (*Cichorium intybus*), Hornklee (*Lotus corniculatus*), Esparsette (*Onobrychis viciifolia*) oder einer Klee / Grasmischung (Kontrollfutter ohne kTs) gefüttert. Anschließend wurden die Tiere zu einer Herde vereint und während 11 Tagen ausschließlich mit Kontrollfutter gefüttert. Diese Periode diente dazu, die Reversibilität einer potentiell durch die Tanninpflanzenfütterung verursachten EpG Reduktion zu untersuchen. Die EpG wurden für alle Vergleiche auf die Kottrockensubstanz (EpG TS) bezogen. Verglichen mit dem Kontrollfutter, führte die Verabreichung aller tanninhaltiger Futterpflanzen Spezies zu signifikant reduzierten EpG TS (Chicorée: um 89 %; Hornklee: um 63 %; Esparsette: um 63 %; alle Tests $P < 0.05$) und einer tendenziellen Reduktion der adulten *H. contortus* (Chicorée: um 15 %; Hornklee: um 49 %; Esparsette: um 35 %). Für keine der tanninhaltigen Pflanzen konnte eine anthelminthische Wirkung gegen *C. curticei* nachgewiesen werden. Das Ausbleiben eines EpG TS Anstiegs während der elftägigen Kontrollfütterungsphase deutet darauf hin, dass die durch die kT-Fütterung verursachte EpG TS Reduktion nachhaltig ist. Die für Hornklee (15.2 g kTs kg^{-1} TS) und Esparsette (26.1 g kTs kg^{-1} TS) gemessenen moderaten bis hohen kT-Konzentrationen stützen die Hypothese, dass kTs für die anthelminthische Wirkung verantwortlich sind. In Bezug auf Chicorée mit einem sehr niedrigen kT-Gehalt (3 g kTs kg^{-1} TS) muss die synergistische oder alternative Wirkung anderer Sekundärmetaboliten in Betracht gezogen werden. Zusammenfassend zeigte dieses Experiment, dass Hornklee und insbesondere Esparsette zur Kontrolle von *H. contortus* beitragen können. Dies sowohl in Bezug auf die Reduktion der Wurmbürde im Tier als auch auf die Kontamination der Weide mit Eiern dieses Parasiten.

Für das zweite Experiment wurden die gleichen MDN Spezies und die gleichen parasitären Stadien (d.h. adulte *H. contortus* und *C. curticei*) wie in Experiment 1 verwendet. Zudem wurde eine vergleichbare Fütterungsperiode gewählt (d.h. 16 Tage). Das Ziel der Studie war es, den anthelminthischen Effekt getrockneter und silierter Esparsette zu untersuchen. Vierundzwanzig parasitennaive Lämmer wurden 28 Tage vor Beginn des Experiments einmalig mit infektiösen Larven dieser Parasiten inokuliert. Die Tiere der Gruppe A wurden mit Esparsettenheu (61 g kTs kg^{-1} TS) und jene der Gruppe B mit Kontrollheu (1 g kTs kg^{-1} TS) gefüttert. Gruppe C erhielt Esparsettensilage (42 g kTs kg^{-1} TS) und den Tieren der Gruppe D wurde Kontrollsilage (1 g kTs kg^{-1} TS) verabreicht. Die entsprechenden Futter wurden ad libitum verabreicht und auf Basis der Futterresten wenn nötig mit Kraftfutter ergänzt, um eine isoproteische und isoenergetische Fütterung zwischen den Gruppen zu gewährleisten. Im Vergleich zu den kontrollgefütterten Tieren reduzierte die Esparsettenfütterung die adulten *H. contortus* (um 47 % bei Heu, P < 0.05; um 49 % bei Silage, P = 0.075), hatte aber einen geringen Effekt auf die adulten *C. curticei*. Verglichen mit der Kontrolle waren die EpG TS der mit Esparsettenheu gefütterten Tiere um 58 % (P < 0.01) und jene mit Esparsettensilage gefütterten Tiere um 48 % (P = 0.075) reduziert. Beide Esparsettenfutter reduzierten die EpG TS von *C. curticei* im Vergleich zur Kontrolle (Heu: um 81 % und Silage: um 74 %, beide Tests P < 0.001). Die Resultate dieses Experiments deuten darauf hin, dass die EpG TS Reduktion von *C. curticei* und *H. contortus* durch unterschiedliche Mechanismen zustande kamen. Im Falle von *H. contortus* scheint diese Reduktion mit einem nematoziden Effekt der Esparsette auf die adulten Würmer zusammenzuhängen. Die EpG TS Reduktion bei *C. curticei* hingegen dürfte eine Folge der reduzierten per capita Fekundität der weiblichen Würmer sein. Da für beide konservierten Esparsettenfutter ein anthelminthischer Effekt gezeigt werden konnte, eröffnen sich viel versprechende Perspektiven für deren Einbezug in eine Kontrollstrategie gegen MDN.

Das Ziel der dritten Studie war es, die anthelminthische Wirkung von Esparsettensilage auf eine gemischte, natürlich erworbene MDN-Population in peri-parturienten Mutterschafen zu untersuchen. Dreiunddreißig Tiere mit EpG > 300 wurden in zwei Gruppen von 16 Tieren (Esparsettengruppe) und 17 Tieren (Kontrollgruppe) aufgeteilt und während 25 Tagen entsprechend gefüttert. Die MDN-Population bestand vorwiegend aus *H. contortus* (~ 50 %) und *Trichostrongylus colubriformis* (~ 35 %). Nach zehn aufeinander folgenden Fütterungstagen waren die EpG TS der Esparsettengruppe im Vergleich zur Kontrollgruppe um 60 % (P < 0.05) reduziert. Obwohl in der Folge nicht mehr signifikant, blieb diese Reduktion konsistent bis zum Schluss des Experiments. Die Fläche unter der EpG TS Kurve (FUK), berechnet für die gesamte experimentelle Periode war in der Esparsettengruppe im Vergleich zu Kontrolle um 32 % (P = 0.17) reduziert. Wurde nur die Phase nach Eintritt der EpG TS Reduktion berücksichtigt (d.h. Tag 10-25), betrug die FUK Reduktion in der Esparsetten Gruppe im Vergleich zur Kontrolle 65 % (P = 0.09). Die Ergebnisse dieses Experiments bestätigen die anthelminthische Wirkung von Esparsettensilage auf eine gemischte MDN-Population in peri-parturienten Schafen. Die Auswirkungen der beobachteten EpG TS Reduktion auf die Epidemiologie der Trichostrongylidose muss allerdings kritisch evaluiert werden.

Zusammenfassend deuten unsere Resultate an, dass der anthelminthische Effekt tanninhaltiger Pflanzen im Labmagen generell bei niedrigerer Futter kT-Konzentration erreicht wird als im Dünndarm. Entsprechend ist der Effekt eher organabhängig als spezifisch für einzelne MDN-Spezies. Im Weiteren konnte gezeigt werden, dass die Konservierung kT-haltiger Pflanzen ohne anthelminthischen Wirksamkeitsverlust möglich ist. Dies bietet interessante Möglichkeiten in Bezug auf eine zentralisierte Produktion, Vermarktung und den erweiterten Einsatz des Futters unabhängig von der Vegetationsperiode. Gegenwärtig genügt die Verfütterung von Tanninhaltiger Pflanzen als alleinige Kontrollstrategie gegen MDN jedoch noch nicht. Die Kombination mit anderen alternativen Kontrollstrategien im Sinne eines integrativen Ansatzes könnte viel versprechend sein, wenn sich die entsprechenden anthelminthischen Effekte ergänzen. Da die Entwicklung und Evaluierung solcher integrativen Kontrollstrategien Gegenstand der laufenden Forschung ist, kann zurzeit nicht auf den Einsatz chemisch-synthetischer Anthelminthika verzichtet werden. Im Hinblick auf die aktuelle Resistenzproblematik der MDN gegen diese Medikamente müssen sie allerdings gezielt und sparsam eingesetzt werden bis geeignete integrative Kontrollstrategien zur Verfügung stehen.

3. Introduction

Sheep harbour a variety of pathogenic gastrointestinal nematodes (GIN). The majority of species belong to the order Strongylida and are further grouped in the super families Trichostrongyloidea, Strongyloidea and Ancylostomatoidea. The focus of the present thesis is on the nematodes residing in the abomasum (*Haemonchus sp.* and *Teladorsagia sp.*) or the small intestine (*Trichostrongylus sp.*, *Cooperia sp.* and *Nematodirus sp.*) and the expression 'GIN' is therefore used synonymously with these species. GIN are ubiquitous and may cause decreases in live-weight gain, fibre and milk production and reproductive performance of small ruminants and therefore seriously impact on animal health and productivity (Sykes, 1994). Comprehensive investigations have concluded that GIN parasitism represents the greatest economic constraint of small ruminant production, whether in the industrialized or the developing world (Perry and Randolph, 1999). Because sheep provide important commodities, maintaining their health and efficiency is crucial to the stability of this resource.

Control of GIN is largely based on preventive or therapeutic use of anthelmintic drugs (Williams, 1997). Frequent use of anthelmintics, however, has led to a rapid development of resistance of parasite populations in sheep in many areas of the world (Jackson and Coop, 2000). This situation is of concern as it is unlikely that many, if any, novel drugs will be licensed for veterinary use in ruminants in the foreseeable future (Geary et al., 1999). Further issues arising from the widespread use of anthelmintics are the increasing public concern about potential drug residues in animal products (Waller, 1999), and the possible ecotoxicological effects of drug excretion on the environment (Bila and Dezotti, 2003). The latter issues, in general, have also provided a stimulus for the rapid expansion of organic agriculture, particularly in Europe, where government subsidies are provided to farmers who opt for this form of production. Statutes developed by various organic farming bodies provide quite rigid stipulations (Thamsborg et al., 1999), and this in particular with respect to the use of synthetic compounds and also to the intense pasture-based livestock management. Therefore, especially in organic farming, difficulties have arisen in maintaining adequate control of pasture-borne infectious diseases such as nematode parasite infections.

These issues have stimulated investigations to find alternative sustainable GIN control strategies that are less reliant on anthelmintic input (Waller and Thamsborg, 2004). Non-chemical approaches to nematode parasite control are at varying stages of development, utility and applicability, and can be considered to be targeted either at the parasitic phases within the animal or at the free-living stages on pastures.

3.1. Alternative approaches to control the free-living stages of GIN

3.1.1. Grazing management
The major epidemiological variable influencing gastrointestinal worm burdens of grazing animals is the infection rate, or the number of infective larvae ingested from pasture each day. In temperate climates, a seasonal pattern of infective larvae is observed which is dependent on the seasonal variation in egg-hatching, larval development and survival, induced by accompanying variation in temperature or rainfall, or of seasonal variation in host susceptibility caused by population events such as parturition or lactation. Knowledge from epidemiological studies therefore permits the implementation of grazing management practices by which the intake of infective larvae by the hosts can be reduced. Although grazing of animals on pastures with a low infectivity combined with targeted anthelmintic treatments have proven highly efficient for cattle, this approach has been less successful in small ruminants (Barger, 1999). Additionally, the anthelmintic treatments usually integrated in grazing management strategies strongly select for anthelmintic resistance in surviving parasites (Le Jambre et al.,

1999), and because of the emergence of anthelmintic resistance, they cannot be relied on as a sole control treatment.

3.1.2. Biological control

To date, biological control of nematode parasites of livestock is almost exclusively associated with the nematode destroying microfungus *Duddingtonia flagrans*. This fungus has the potential to break the GIN life cycle by capturing larval stages in the faeces before they migrate to pasture. The fungus has the properties to (i) survive the gut passage of ruminants, to (ii) grow rapidly in freshly deposited faeces and (iii) voraciously consume nematodes (Larsen, 1999). Practically, spores of the fungus have to be administered to the animal as a feed additive. Field evaluations of this approach for a range of livestock species have been under-way for the last decade and have proven the effectiveness of the concept in cattle and to a lesser extent in small ruminants (Larsen et al., 1995, Faedo et al., 1997, Larsen et al., 1998, Eysker et al., 2006). The reduced effectiveness in sheep and goats is mainly attributable to the structure of sheep and goat faeces leading to a poorer growth of the fungus. In order to achieve optimal results, the spores need to be continuously shed in the faeces concurrently with output of parasite eggs. Thus, daily supplementation of spore material is recommended at a period predetermined by the epidemiological background in order to produce satisfactory effects. To date, the route of administration and the necessary period of administration (e.g. 3 months) clearly represent serious drawbacks in the practicability of the approach. Also, the development of controlled release devices or feed-blocks has not yet reached a sufficient level of effectiveness and practicability.

3.2. Alternative approaches to control GIN within the host

3.2.1. Genetic resistance

Genetic resistance of the host would be the ideal for a sustained control of parasites. It represents a low-cost, permanent solution requiring no additional resources (Waller and Thamsborg, 2004). It has been shown repeatedly that there is substantial variation in host resistance to GIN and that some of the variation is due to genetic factors of the host (Stear and Murray, 1994). There are, however, major differences in genetic resistance between different breeds, as a general rule being low in high productivity breeds and being higher in more extensive breeds. Ruminants highly resistant to GIN infection are mainly found in the tropics, where the combination of malnutrition, environmental stress, the often massive larval challenge and limited use of anthelmintics have imposed the harshest conditions for the selection of the fittest animals (Baker, 1998). However, the accumulated data from studies with such GIN resistant animals suggest a negative correlation between host resistance and productivity traits such as wool or meat yield (Stear and Murray, 1994). There are attempts to identify genes that encode parasite resistance in laboratory animal models and to identify the locations of similar genes in ruminants with the aid of genomic maps (Beh and Maddox, 1996, Gray and Gill, 1993, Sonstegard and Gasbarre, 2001). Irrespective of the benefits derived from the use of host genetics to control nematode infections, progress is likely to be slow because of the long generation interval of livestock and if transgenic animals are to be produced, there is likely to be a considerable debate.

3.2.2. Vaccines against GIN

To date, vaccines against nematode parasites have had very little commercial success. Early experiments in this field have been undertaken using attenuated whole parasites, but with the exception of the cattle lung worm (*Dictyocaulus viviparus*), this approach failed, in particular for the economically important GIN (Bain, 1999). Attention was then focused on recombinant vaccines, using antigens thought to be protective against helminth parasites. Although these efforts led to some successes, particularly with the abomasal nematode Haemonchus contortus, major barriers still exist for most other GIN species. The sources of these complications are manifold and include the low protection level to artificially produced antigens, strategies of the parasites to evade immunological effector mechanisms and also the complexity of the host's immune response which seems to involve a combination of local hypersensitivity, cell-mediated, antibody and inflammatory responses (Smith, 1999). Lastly, the natural unresponsiveness of young animals (< 3-6 month of age) against some GIN, and in female animals around parturition further complicate the situation (Knox, 2000)

3.2.3. Herbal anthelmintics

The origins of anthelmintic medication are found in the use of various plant preparations. These traditional health practices rapidly disappeared from the human and veterinary market with the discovery of synthetic anthelmintic compounds (Waller et al., 2001). However, in developing countries or in regions where the access to synthetic drugs is limited by cost or due to supply, a diverse range of herbal de-wormers is still frequently used for treatment of GIN and other types of parasites. In general, there is a lack of scientific validation of the purported anthelmintic effect of these products (Schillhorn Van Veen, 1997, Githiori et al., 2006), although this field represents an area with some potential for future anthelmintic solutions.

In recent years, the interest in traditional health practices, including herbal de-wormers has experienced renewed attention in both, the industrialized and developing countries of the world. With respect to GIN, studies are now under-way to evaluate some of these candidates traditionally used as livestock de-worming preparations from East Africa and Asia but also from more temperate regions (Hördegen et al., 2003, Githiori et al., 2003, Githiori et al., 2004). A particular group of compounds — the cystein proteinases — has been suggested to have high potential as a novel group of anthelmintics, as they have been shown to damage the cuticle of nematodes (Stepek et al., 2004). However, their mode of action is not very specific and the safety index (maximum tolerated dose / recommended therapeutic dose) is expected to be low. With respect to low-input farming, the role of herbal anthelmintics is difficult to envisage, because marketing of products with a high level of efficacy will inevitably be accompanied by regulatory requirements addressing residue levels and human safety and therefore will become quasi-natural and regarded as medical treatment.

3.2.4. GIN and host nutrition

GIN impair animal productivity through reductions in voluntary feed intake and / or reductions in the efficiency of food use (see introduction), particularly by inefficient absorption of nutrients in the gastrointestinal tract (Coop and Kyriazakis, 2001). This is most pronounced for protein, for which a reduced absorption and / or retention in parasitized animals has repeatedly been shown (Coop and Kyriazakis, 1999) but also the reduced absorption of minerals, especially phosphorus, is also of high significance (Poppi et al., 1985). The magnitude of these effects is influenced by the size of larval challenge and the number and species of worms present. A common feature of many GIN infections is an increased loss of endogenous protein into the gastrointestinal tract, which is partly attributable to increased leakage of plasma protein, increased sloughing of epithelial cells and increased secretion of mucoproteins (Van Houtert and Sykes, 1996). Dependent on whether the lesions induced by the GIN are in the anterior or the distal part of the gastrointestinal tract, some of the protein is reabsorbed if there is adequate compensatory absorptive capacity (Coop and Holmes, 1996). Despite some re-absorption, protein losses are large and generally more pronounced for infections of the small intestine. The gastrointestinal tract is a highly competitive tissue and, when its demand for nitrogen is increased by parasitism, such as during repair processes, the consequence will be a net loss of nitrogen in other tissues, thus affecting growth and the reproductive effort (Symons and Jones, 1975). These effects will be exacerbated by the fact that nutrient availability is further limited by the reduced feed intake. Lambs and lactating females are the most susceptible to nematode infections (Barger, 1993). Strategic feed supplementation, particularly to infected young and periparturient animals, can thus have short and even long-term benefits, and research is now targeted at fine-tuning of supplement administration.

A further approach to influence the development and consequences of parasitism by ways of nutrition is represented by the intake of antiparasitic compounds naturally present in the fodder. Various secondary plant metabolites such as nitrogen containing metabolites, terpenoids and also phenolic metabolites are thought to have antiparasitic properties (Anthony et al., 2005). Within the structural group of phenolic metabolites, extensive research effort has been dedicated to the anthelmintic activity of condensed tannins in recent years (Hoste et al., 2006).

3.3. Tannins

3.3.1. Biochemical structures and properties of tannins
Tannins are secondary plant polyphenols with great diversity. They are produced by a wide range of plant species with some variability with respect to the plant organ. Tannins are functionally defined by their capacity to bind proteins, which was traditionally exploited by the leather industry to preserve (tan) leather (Mueller Harvey and McAllan, 1992). Tannins are categorized into two major structural groups: i) hydrolysable and ii) condensed tannins. Hydrolysable tannins (HTs) are gallic or ellagic acid esters of sugars (Mueller Harvey, 2001). When they are consumed by ruminants, they can be degraded into gallic acid, which is readily absorbed from both the rumen and the small intestine and has been associated with liver and kidney lesions in sheep (Zhu *et al.*, 1992).
Condensed tannins (CTs) are polyphenolic oligomers and polymers of catechin (flavan-3-ols). The depolymerisation products of CTs are cyanidin and prodelphinidin and CTs have therefore been further classified as procyanidins and prodelphinidins (Waterman, 1999). Only a low degree of absorption of CTs by the digestive tract of ruminants has been reported (Terrill *et al.*, 1994). One of their most important chemical properties is the ability to form soluble and insoluble complexes with macromolecules, such as protein, fibre and starch (Luck *et al.*, 1994, Haslam, 1996). The CT-protein interactions are most frequently based on hydrophobic and hydrogen bonding and are determined by the molecular mass and the molecular structure of both the tannin and the protein. Tannin-protein binding is usually reversible and acidic or alkaline pH or treatment with organic solvents can result in the dissociation of the complexes. For example, Jones and Mangan (1977) reported that CTs can bind with protein at pH 3.5-7.5 (i.e., ruminal and small intestinal conditions) to form CT-protein complexes, which dissociate and release protein at pH below 3.5 (i.e., abomasal conditions) or above 7.5.

3.3.2. Tannins in ruminant fodder plants
HTs are present in a range of trees such as Oak (*Quercus spp.*), Acacia (*Acacia spp.*) or Eucalyptus (*Eucalyptus spp.*), but also in a range of other shrub and tree leaves. These leaves can form a major component of ruminant diets in mediterranean and tropical regions where grasses are of poorer quality and lower availability. Under these conditions, ruminants become increasingly dependent on browse as a food and N source (Waghorn and McNabb, 2003). The leaves and apices of these plants can contain up to about 200 g HT kg^{-1} DM. In contrast, condensed tannins are the prevailing type of tannins in temperate legumes, with concentrations up to 200 g kg^{-1} DM (Reed, 1995). Legumes such as birdsfoot trefoil (*Lotus corniculatus*), big trefoil (*Lotus pedunculatus*), sainfoin (*Onobrychis viciifolia*), sulla (*Hedysarium coronarium*) but also plants from other families like Sericea lespedeza (*Lespedeza cuneata*) are known to express significant quantities of CTs in their foliage (Mosjidis *et al.*, 1990, Terrill *et al.*, 1992). These forages perform well under conditions of average or poor soil fertility and some are tolerant to acid soils (e.g. lotuses). CTs are usually absent or have very low concentrations in the leaves of grasses. In browse from tropical legume species in particular, the CTs are often accompanied by a wide range of secondary metabolites, including toxins such as mimosine (Lowry *et al.*, 1996).
Generally, expression of tannins (HTs and CTs) seems to depend to a considerable extent on abiotic factors such as light, temperature and rainfall and also biotic factors such as nutrient availability and herbivore pressure to the plant (Waterman, 1999). Additionally, the CT-concentrations measured during the life cycle of the plant are variable, with the CT-content usually being highest in young, pre-flowering plants and lowest in the reproductive stages (Häring *et al.*, 2007). This is also illustrated in Table 4.1, where the CT-levels of the same plant species vary considerably.
CTs are confined to cell vacuoles and are essentially unreactive until released by cell rupture, for example by chewing, resulting in extensive binding with proteins (e.g. plant, salivary, animal, microbial, enzymes etc.; Waghorn and McNabb, 2003). Unlike the location of CTs within the plant,

HTs are deposited in mesophyll cell walls with no evidence of storage in vacuoles (Grundhofer et al., 2001).

3.3.3. Nutritional and animal production aspects of tanniferous plants in non-infected ruminants
Because CTs are the predominant type of tannins in temperate fodder plants, substantial research has been dedicated to their effect on ruminant nutrition. With respect to this subject, they are of interest because of their reactivity with forage proteins after the plant has been chewed, resulting in the formation of protein-CT complexes in place of the 'normal' solubilisation (Barry and McNabb, 1999). Aside from reductions in protein solubility and degradation rate by rumen microbes, the mechanisms by which CTs behave in the rumen are poorly understood. The consequences, however, are clear: i) an increased outflow of undegraded plant protein (bypass protein) to the intestines and ii) substantial reductions in rumen NH_3; the extent of these processes being largely dependent on the tanniferous plant species and its CT-concentration (McNabb et al., 1996, Waghorn and McNabb, 2003). The consumption of tanniferous plants with high concentrations of CTs (> 60 g kg^{-1} DM) has been associated with a number of detrimental effects on the metabolic nutrient supply and performance of ruminants, including reduction in voluntary feed intake, growth inhibition and interference with the morphology and the proteolytic activity of microbes in the rumen or depressed growth rates. Low or moderate CT-concentrations (< 60 g kg^{-1} DM) have resulted in positive effects on sheep (Min et al., 2003a). For example, increased live weight gain, reduced carcass fat content as well as increased milk and wool production have all been associated with the consumption of CTs (Waghorn and McNabb, 2003).
It is generally accepted that the beneficial effects observed following the consumption of tanniferous plants are mainly due to the increased availability of non-degraded protein in the small intestine (Min et al., 2003a, Waghorn and McNabb, 2003). Currently, research is focused on other beneficial effects of CTs on livestock production, for example the reduction of the detrimental effects of parasitism in grazing small ruminants.

3.3.4. Antiparasitic and physiological effects of tanniferous plants in infected ruminants
The majority of published work on the anthelmintic effect of tanniferous fodder plants has focused on forage legumes (*Fabaceae*), including sainfoin (*Onobrychis viciifolia*) (Paolini et al., 2005b, 2003b, 2005a, Hoste et al., 2005, Athanasiadou et al., 2005, Thamsborg et al., 2004), sulla (*Hedysarium coronarium*) (Niezen et al., 1995, 1998a, 2002a, Tzamaloukas et al., 2005, Athanasiadou et al., 2005, Pomroy and Adlington, 2006), Sericea lespedeza (*Lespedeza cuneata*) (Lange et al., 2006, Shaik et al., 2006, Min et al., 2004, 2005), birdsfoot trefoil (*Lotus corniculatus*) (Bernes et al., 2000, Marley et al., 2003b, Niezen et al., 1998a) and big trefoil (*Lotus pedunculatus*) (Niezen et al., 1998a, 1998b, Tzamaloukas et al., 2005, Athanasiadou et al., 2005). Some studies also investigated the anthelmintic effect of chicory (*Cichorium intybus*) from the Asteracea family (Thamsborg et al., 2004, Scales et al., 1994, Athanasiadou et al., 2005, Tzamaloukas et al., 2005, Marley et al., 2003b). Because of it's very low CT-content (usually < 5 g CTs kg^{-1} DM), chicory can not be considered to be a tanniferous plant. However, as chicory has been included in various studies together with CT-rich plants, it is mentioned in this thesis for the sake of completeness. The anthelmintic effect of chicory is believed to be associated with other secondary plant metabolites (see section 3.3.6. & 5.4.2.).

Beneficial effects of tanniferous plants on the host physiology and performance of parasitized small ruminants have generally been observed in studies where the consumption of a CT-containing plant was compared to a similar CT-free control fodder (Table 4.1). This has been evaluated by (i) comparison of the clinical status of animals (ii) pathophysiological measurements and (iii) assessment of the impact of parasitism on production when infected sheep, goats or deer consume legumes or chicory (Table 4.1). For example, lambs artificially infected with *Trichostrongylus colubriformis* and *Teladorsagia circumcincta* had higher daily live weight gains when grazing big trefoil (*Lotus pedunculatus*;

56 g CTs kg^{-1} DM), compared to perennial ryegrass (*Lolium perenne*), which does not contain significant amounts of CTs (Niezen *et al.*, 1998b).

In addition to the favourable effects on host physiology and the ability to maintain homeostasis despite parasite infections, the consumption of tanniferous plants has also been associated with antiparasitic effects (Table 4.1). For instance, in the study cited above, apart from the increased weight gain of animals fed big trefoil, faecal egg count (FEC) and worm burden of *T. circumcincta* where also significantly reduced when compared to the control (Niezen *et al.*, 1998b). Overall, the most commonly reported antiparasitic effect has been a substantial decrease in FEC. This can contribute towards reducing contamination of pastures with larvae and consequently lowering animal infections while grazing. Data on reductions of female worm fecundity or worm number have also been reported, but, compared to FEC reductions, appear to be less consistent. Some researchers suggested that the development of eggs to larvae in the faeces might be impaired in animals consuming tanniferous fodder (Marley *et al.*, 2003a, Niezen *et al.*, 1998b, 2002b). Finally, lower numbers of infective larval stages on the plant stratum grazed by ruminants where observed for some tanniferous forages swards compared to swards of other forages such as white or red clover, potentially reducing infections of the host (Moss and Vlassoff, 1993). Generally, however, the antiparasitic effects reported for both the parasitic stages and the free-living stages of GIN seem variable, depending on the tanniferous forage plant used, the environment in which the plant was grown, the specific stage and composition of nematode population and possibly also on the host species (Table 4.1).

3.3.5. Mechanism of CT-action

Two hypothesis have been put forward to explain the anthelmintic efficacy of CT-rich fodder plants.

3.3.5.1. Indirect effects

As mentioned in the preceding section, CTs can protect proteins from degradation in the rumen and therefore increase the protein flow to, and amino acid absorption by the small intestine. As any increase in intestinal protein supply is known to improve host homeostasis and its immune response against GIN, the improved utilisation of nutrients by hosts receiving moderate amounts of CTs could thus contribute to the improvement in resilience and also modulate host resistance (Coop and Kyriazakis, 2001). This hypothesis is often referred to as the 'indirect mode of CT-action'. With respect to this hypothesis only few studies have been performed where local or general parameters of host immunity have been recorded, and the results remain largely inconclusive (Athanasiadou *et al.*, 2005, Tzamaloukas *et al.*, 2005, Niezen *et al.*, 2002a, Paolini *et al.*, 2003c, Athanasiadou *et al.*, 2001b). It is possible, however, that this route of CT-action might contribute substantially to the action observed when CT-rich fodder is administered to GIN infected sheep or goats.

3.3.5.2. Direct effects

The second hypothesis of CT-action is based upon the possibility that CTs could have intrinsic anthelmintic properties that directly affect biological processes of the adult and larval worm-stages. The investigation of this 'direct mode of CT-action' has been the subject of several in vivo investigations in sheep and goats using quebracho as a source of CTs (Table 4.2). Quebracho is an extract of the bark of the tropical tree *Schinopsis sp.* with a CT-content of 600 – 700 g kg^{-1} DM. With the exception of one study (Athanasiadou *et al.*, 2000b), the short term design of these experiments did not allow the development and expression of effective host immune responses. Additionally, the experimental feeds of the CT-fed group and the control fed animals were isonitrogenous and isoenergetic in order to rule out any indirect effects. In general, the addition of quebracho to the feed of parasitized sheep or goats resulted in either a decrease in the establishment of infective larvae (Paolini *et al.*, 2005b, 2003a) or reductions in worm fertility and egg output (Athanasiadou *et al.*, 2000b; Table 4.2). Interestingly, in these experiments with quebracho, the variability of the effect seen in experiments with

CT-rich fodder plants was to some extent also observed.

Other researchers aimed at showing the existence of direct anthelmintic effects of CTs in experiments where polyethylene glycol (PEG) was added to the experimental CT-containing fodder. PEG is known from nutritional experiments with sheep and goats to effectively bind CTs and to inactivate them. These studies with parasitized animals have given contrasting results. The addition of PEG to goats browsing on tanniferous forage significantly increased the nematode egg excretion (Kabasa et al., 2000) whereas in parasitized sheep grazing on big trefoil, the addition of PEG did not result in significant parasitological differences compared to an un-supplemented group grazing the same forage (Niezen et al., 1998b). In the latter case, it is not clear why the addition of PEG to big trefoil did not influence the antiparasitic effects.

The role of CTs as an antiparasitic agent has also been extensively studied in vitro. In vitro assays target either the eggs (egg hatch assay, EHA), the feeding of larval stages (larval feeding inhibition assay, LFIA), the larval development (larval development assay, LDA) or the motility of infective larvae (larval migration inhibition assay, LMIA). All these assays have been used routinely to evaluate the effects of crude extracts or specifically extracted CTs of tanniferous plants.

The addition of quebracho or whole plant extracts from chicory, sulla or sainfoin to the in vitro culture medium has usually led to significant dose-dependent inhibitory effects on the functions evaluated with the respective assays (Molan et al., 2003b, 2003a, 2000c, Paolini et al., 2004). Moreover, in some of these experiments the inhibitory properties of PEG on CTs where exploited. The addition of PEG to the assays usually led to restoration to control values (Molan et al., 2000c, Paolini et al., 2004). Some of the most convincing evidence supporting an antiparasitic role of CTs has been obtained from activity-guided in vitro assays comparing the activity of specific biochemical fractions of sainfoin obtained by various extraction procedures. The influence on nematode motility as measured by the LMIA was mainly associated with the fractions containing CTs (Barrau et al., 2005).

It has been shown by scanning electron microscopy that the in vitro incubation of adult *T. colubriformis* with CT-extracts has led to cuticular changes of the worm surface (Hoste et al., 2006). These changes could be related to the binding of CT-molecules to parasite surface proteins and might also explain the results from a recent in vitro experiment in which CT-extracts inhibited the exsheathment of third-stage larvae (Bahaud et al., 2006). However, the exact mechanism of CT-action still remains largely unclear and could depend on several parameters related to the plant, the parasite species and its specific stage and also the host species.

Table 4.1. Summary of results of all in vivo experiments performed with Sainfoin (*Onobrychis viciifolia*), Sulla (*Hedysarum coronarium*), Big trefoil (*Lotus pedunculatus*), Birdsfoot trefoil (*Lotus corniculatus*) and Chicory (*Cichorium intybus*) in sheep and goats.

Plant species	Host species	Experimental design and exposure	CT-content (g CT kg⁻¹ DM)	Infection type/GI nematode species	FEC reduction % [a]	Worm reduction % [a]	PCF/Fecundity reduction [a]	Productivity/Pathophysiology [a]	Reference
Sainfoin (*Onobrychis viciifolia*)	Goat	Regular distribution of hay 1 week / month for 3 months	27	NI mainly *H. contortus* (39 %) and *T. colubriformis* (42 %)	~ 65 % sign.	*T. colubriformis* 50 %, *H. contortus*, no difference	Eggs in utero sign., both GIN species	PCV sign. elevated	A
	Goat	Regular distribution of hay 1 week / month for 9 months	25	NI mainly *T. circumcincta*, and *T. colubriformis*	Lower over 9-month period sign.	NE	NE	No difference in milk production	B
	Sheep	2 infections, days 0 and 35 of experiment, grazing from day 28 – 42 (i.e. 14 days)	14.9	EI *T. colubriformis*	NE	NE	NE	NE	C
	Sheep	10 weeks experiment; 2 infections, at days 0 and 28 of experiment, Evaluation of effect on established and incoming worms	41	EI *T. circumcincta* and *T. vitrinus*	80 % for established worms, 40 % for incoming worms	*T. circumcincta* 35 % n.s.	NE	NE	D
	Goat	Experiment on establishment, single infect. on 3 consecutive days, then 9 days of hay feeding	32	EI *H. contortus*	NE	38.5 % n.s.	NE	No difference in PCV and pepsinogen concentrations	E
	Goat	Hay feeding for 20 days	32	NI	~ 50 % sign.	NE	NE	NE	F
Sulla (*Hedysarum coronarium*)	Goat	10 days indoors experiment	45	mainly *T. circumcincta* 81 % and *T. vitrinis* 15 %	NE	No difference	NE	NE	S
	Sheep	a) 28 days grazing b) 42 days grazing	100-120	a) NI b) EI *T.colubriformis*	a) ~ 40 % sign. b) Lower over 42 days period sign.	a) NE b) ~ 50 % sign.	a) NE b) NE	a) increased LWG b) increased LWG and wool production	G
	Sheep	42 days grazing experiment	ND	NI mainly *T.colubriformis* (63 %)	Lower over 42 days period sign.	30 % n.s.	NE	Increased LWG, No difference in wool growth	H
	Sheep	~ 56 days grazing experiment	31	EI *T. colubriformis / T. circumcincta*	~ 43 % sign.	*T. circumcincta* 48 % sign., *T. colubriformis* ND	NE	NE	I
	Sheep	2 infections, days 0 and 35 of experiment, grazing from day 28 – 42 (i.e. 14 days)	15.8	EI *T. colubriformis*	No difference	No difference	No difference	No difference in LWG	C
	Sheep	35 days experiment Infections on days 0 and 28, grazing from day 21-35	15.8	EI *T. circumcincta*	No difference	No difference	No difference	NE	J
Sericea lespedeza (*Lespedeza cuneata*)	Sheep	a & b) 49 days of SL hay feeding, day 0-21 trickle infect. 3x/week	224	a) Existing NI and EI with *H. contortus* b) EI *H. contortus*	a) Lower in the course of the exp., sign. b) Lower in the course of the exp., sign.	a) 67 % n.s. (only 3 animals per group killed b) 26 % red. n.s.	NE	a) PCV sign. elevated b) No difference	K
	Goat	Cross over design! 15 days Sericea grazing then 15 days control and	46 (extractable CT only)	NI mainly *H. contortus* (91 %)	75 % sign.	NE	NE	NE	L

18

						circumcincta 26 % T. colubriformis 40 %, all sign.			
	Goat	81 days of SL grazing	152	NI	80 %	H.contortus 89 % T. circum-cincta, 100 % T. colubriformis 50 %, all sign.	NE	No difference in milk compo-sition, tendency of increa-sed LWG (P=0.1)	N
Birdsfoot trefoil (Lotus corniculatus)	Sheep	42 days grazing experiment	ND	NI mainly T.colubriformis (63 %)	No difference	No difference	NE	No difference in LWG	H
	Sheep	35 days grazing experiment	ND	NI	Day 7: 53 % sign., day 35: ND	32 % sign.	NE	No difference in LWG and BCS	O
	Sheep	14 days of trickle infect., experi-ment on established burden 42 days of cut fodder administration	8	Exp. Trickle infection of natural isolate	No difference	No difference	NE	NE	P
Big trefoil (Lotus pe-dunculatus)	Sheep	42 days grazing experiment	ND	NI mainly T.colubriformis (63 %)	No difference	No difference	NE	Increased LWG, no difference in wool growth	H
	Sheep	42 days exp. feeding, infect. 7 days post onset of feed.	56	NI	56 % sign.	T. circumcincta 23 % sign., no effect on T. colu-briformis	No difference	Increased LWG	Q
	Sheep	35 days experiment Infections on days 0 and 28, grazing from day 21-35	16	EI T. circumcincta	No difference	No difference	No difference	NE	J
	Sheep	2 infections, days 0 and 35 of ex-periment, grazing from day 28 – 42 (i.e. 14 days)	15.9	EI T. colubriformis	No difference	No difference	No difference	No difference in LWG	C
Chicory (Cichorium intybus)	Sheep	10 weeks experiment; 2 infections, days 0 and 28 of experiment, Evaluation of effect on established and incoming worms	11.6	EI T. circumcincta and T. vitrinus	No difference	T. circumcincta 55 % n.s.	NE	NE	D
	Sheep	Single infection on day 0, then 64 days grazing	ND	NI mainly T. circumcincta (70 %) and T. colubriformis (25 %)	No difference	T. circum-cincta 62 % sign., no effect on T. colu-briformis	NE	Increased LWG	R
	Sheep	2 infections, days 0 and 35 of ex-periment, grazing from day 28 – 42 (i.e. 14 days)	< 5	EI T. colubriformis	No difference	33 % n.s.	No difference	No difference in LWG	C
	Sheep	35 days experiment Infections on days 0 and 28, grazing from day 21-35	< 5	EI T. circumcincta	No difference	T. circumcincta 43 % adult burden n.s.	No difference	NE	J
	Sheep	35 days grazing experiment	ND	NI	No difference	Abomasal burden 23 %	NE	Increased LWG and BCS	O

[a] Compared to non-CT-fed controls

Abbreviations: EI, experimental infection, NI, natural infection; ND, not determined; NE, not evaluated; LWG, live weight gain; BSC, Body condition score; PCV, packed cell volume; sign., significant; n.s., not significant;

References: A) Paolini et. al. (2005a); B) Hoste et. al. (2005); C) Athanasiadou et. al. (2005); D) Thamsborg et. al. (2004); E) Paolini et. al. (2005b); F) Paolini et. al. (2003b); G) Niezen et. al. (1995); H) Niezen et. al. (1998a); I) Niezen et. al. (2002a); J) Tzamaloukas et. al. (2005); K) Lange et. al. (2006); L) Min et. al. (2004); M) Shaik et. al. (2006;)N) Min et. al. (2005); O) Marley et. al. (2003b); P) Bernes et. al. (2000); Q) Niezen et. al. (1998b); R) Scales et. al. (1995); S) Pomroy et. al. (2006)

3.3.6. Variability of anthelmintic effects of CTs

Most studies investigating the effects of tanniferous forages against GIN of sheep and goats usually compared the experimental CT-rich fodder to a control fodder not containing CTs but differing in many other aspects (e.g. protein or energy content). In many cases it is therefore difficult to exclude the involvement of indirect CT-action. The potential implication of both, direct and indirect effects in these studies therefore might account for some of the observed variability in results. However, irrespective of the mechanism(s) involved in causing the effects, the accumulated data on anthelmintic effects of several tanniferous fodder plants suggests that (i) the administered CT-dose, (ii) the chemical structure of the CTs, (iii) the exposure of GIN to CTs and (iv) the potential involvement of other secondary plant metabolites are major factors modulating their activity against nematodes.

CT-dose

With the exception of chicory, a threshold of at least 20-30 g CTs kg^{-1} DM must be reached to observe anthelmintic effects (Table 1). On the other hand, little work has been done to determine whether higher doses (> 70 g CTs kg^{-1} CT) substantially increase the antiparasitic effect. From an animal nutrition point of view, however, as mentioned in section 4.3., such CT-levels were frequently associated with low voluntary feed-intake as well as disturbances of the digestive physiology, which tend to outweigh the positive anthelmintic effects (Athanasiadou et al., 2001a, 2000b). Interestingly, recent studies with goat and sheep reported strong antiparasitic effects for Sericea lespedeza (*Lespedeza cuneata*), comprising high levels of CTs (150-230 g kg^{-1} DM) without any negative effect on voluntary feed intake or performance of the host (Min et al., 2003b, Lange et al., 2006). In general, however, when the CT-dose is increased for antiparasitic purposes, the possible anti-nutritional consequences must also be taken into account.

CT-structure

In addition to the importance of the CT-concentration in the diet, experimental evidence from in vitro studies indicates that the molecular structure of CTs may be an important factor influencing the antiparasitic activity. Experiments using only prodelphinidins (one form of the CT monomers) found a stronger in vitro activity against GIN than for procyanidins, suggesting tanniferous plants with a high prodelphinidin to procyanidins ratio to be more effective against GIN (Molan et al., 2003b). This is partly supported from a comparison of in vivo experiments, where the highest level of anthelmintic activity was observed in plants with particularly high prodelphinidin to procyanidins ratios (Min and Hart, 2002).

Availability of CTs in the host and exposure to nematodes

Differences of the CT-activity in vivo could be also explained by differences in exposure of nematodes to CTs depending on the local environment. Formation of CT-protein bonds is a complex phenomenon which is influenced by the structure and the molecular weight of both the CTs and the proteins as well as the biochemical environment (e.g. pH) (Mueller Harvey and McAllan, 1992, Haslam, 1996). With respect to the ruminant digestive tract the near neutral pH in the rumen is favourable for the formation of CT-protein complexes whereas the acidic conditions in the abomasum may lead to their dissociation. Despite their potential dissociation from other proteins in the abomasum, the re-association of free CTs (unbound CTs) with the proline-rich nematode cuticle proteins seems possible, as CTs are known to have a particular affinity to proline-rich proteins (Hoste et al., 2006). With respect to the small intestine, experiments with uninfected sheep have shown that CTs in this environment are principally bound to protein under the prevailing near neutral pH conditions (Waghorn and McNabb, 2003). It is therefore less reasonable to relate the observed anthelmintic effects on intestinal GIN to the action of free CTs. Although little is known on the molecular interactions and the reactivity of CTs in the different parts of the gastrointestinal tract, the nematode stage and species differences in response to CTs might well relate to differences in the availability of CTs and also to the actual time of exposure of worms to CTs.

Variation between hosts in response to tanniferous plant administration might also be related to differences between the conditions in the gut of the hosts. In contrast to sheep, it has been shown that goats are physiologically adapted to counteract secondary plant metabolites present in browse, which could modulate the effect (Silanikove *et al.*, 1996).

Other secondary metabolites of CT-rich plants

It has been shown that some of the tanniferous forage plants tested for anthelmintic activity in the past in fact contain additional secondary metabolites which have shown anthelmintic properties in vitro. This is particularly pronounced for chicory samples containing less than 5 g CTs kg^{-1} DM but large amounts of other phenolic metabolites potentially active against GIN (Tzamaloukas *et al.*, 2005) as well as sesquiterpene lactones that have shown anthelmintic efficacy in vitro (Molan *et al.*, 2003a, 2000b). Recently, Barrau *et al.* (2005) showed that sainfoin, apart from CTs, also contains the anthelmintic active flavonol glycosides: rutin, narcissin and nicotiflorin. The presence of these additional active compounds, competing or interacting with CTs might account for some of the variability recorded in various in vivo studies.

4. Goals of current research

Although a substantial body of evidence has been accumulated with respect to anthelmintic effects of tanniferous fodder plants, research in this area can still be considered to be in its early stages as it is largely focused on the screening and identification of CT-rich fodder plants that have demonstrable activity against GIN. Overall, this screening provides evidence for anthelmintic effects associated with tanniferous plants and the important role of CTs in producing these effects has been shown. However, they also high-light the large variability of anthelmintic effects in response to the consumption of tanniferous fodder plants. The factors potentially contributing to the observed variability have been reviewed in the preceding sections and the exploration of the background of this variability certainly represents a major challenge in this field of research.

Some key questions and issues that need to be addressed include:

A) Direct anthelmintic effects have only been demonstrated with quebracho CTs. Are they also demonstrable with tanniferous fodder plants?
B) Are the differences in effect between different GIN species explicable in terms of differences in bioavailability of CTs in different parts of the gastrointestinal tract or are they related to differences in susceptibility between parasites?
C) The majority of the existing studies focused on effects of tanniferous plants against the intestinal GIN species *Trichostrongylus colubriformis* and the abomasal species *Teladorsagia circumcincta*. In order to answer question C, information on other GIN is urgently needed.
D) The most coherent anthelmintic effect reported from in vivo studies is a reduction in faecal egg count. Is this effect produced by temporally reduced fecundities of female worms and therefore potentially reversible?
E) Given that CT-concentration is an important factor determining the anthelmintic effect, is it possible to establish an in vivo dose-response relation between the CT-dose and the parasitological effect?
F) What is the best way of exploiting tanniferous forages for anthelmintic means – in their native state as forages or as cut and conserved products (i.e. silage or hay)?
G) Given the current state of research, what are the possibilities to implement production and utilisation of tanniferous forages in livestock producing farms?
H) What is the mechanism of anthelmintic action of CTs and other secondary metabolites present in tanniferous plants?

This thesis reports three experiments that were conducted in order to address most of these questions and issues raised above.
The following chapters contain:

Experiment 1 – addressing issues A, B, C, D, E
The objective of this experiment was to test the anthelmintic effect of freshly administered sainfoin, birdsfoot trefoil and chicory against *Haemonchus contortus* and *Cooperia curticei* under controlled conditions.
Experiment 2 – addressing issues A, B, C, D, F
The aim of the second experiment was to determine whether anthelmintic activity on *Haemonchus contortus* and *Cooperia curticei* is maintained when using conserved (hay and silage) tanniferous plant material (sainfoin).
Experiment 3 – addressing issues F, G

In the third study the anthelmintic effect of ensiled sainfoin was examined when administered to peri-parturient ewes carrying a mixed GIN population in a on-farm setting.

This work is part of an interdisciplinary Swiss project on tanniferous fodder plants between the Swiss Federal Research Institutes ALP Posieux (Animal Science and Nutrition) and ART Reckenholz (Agroecology and Plant Sciences), as well as the Research Institute of Organic Agriculture (FiBL — Parasitology). Within this framework, in addition to parasitological aspects considered here, the major research topics of the other partners include the nutritional properties of tanniferous forages, also considering potential anti-nutritional effects (ALP Posieux) and the cultivation of tanniferous plants, addressing aspects of yield, concurrence and CT-expression during the life-cycle of the plant (ART Reckenholz). The research of the latter two parts of the project is still ongoing.

5. Individual administration of three tanniferous forage plants to lambs artificially infected with *Haemonchus contortus* and *Cooperia curticei*

Based on: Heckendorn F., Häring D.A., Maurer V., Senn, M., Hertzberg, H.
(accepted for publication in: Vet. Parasitol., January 15th 2007)

5.1. Introduction

The wide spread development of anthelmintic resistant populations of gastro-intestinal nematodes (GIN) during the last decade (Jackson and Coop, 2000) and the requirements for reduced anthelmintic input in organic production systems (Githiori *et al.*, 2004) necessitates the development of alternative, non chemical control strategies against GIN. Recent research has suggested that forages containing CTs may offer a promising alternative approach for the control of GIN. In several experiments, the consumption of tanniferous forages was associated with reduced levels of GIN parasites and improved performance of small ruminants (Min and Hart, 2002). It has been hypothesised that the effect of CTs against GIN might be indirect, by long term improvement of host immunity due to an increased protein availability or direct, by short term affection of several key biological processes of parasites (Hoste *et al.*, 2006). In vitro and in vivo experiments with sheep and goats in which the short-term experimental design did not permit the development and expression of host immune responses support the hypothesis of a direct effect of CTs against GIN (Molan *et al.*, 2000c, 2001, Paolini *et al.*, 2004, 2003a, Heckendorn *et al.*, 2006). These studies, however, also pointed out that the anthelmintic effect of tanniferous forages is variable, depending on the GIN parasite species, the parasitic stage, the CT-containing forage plant species and probably also on the host species used. For example, lambs carrying adult *Teladorsagia circumcincta* worms and fed on a fresh CT-forage, such as sainfoin (*Onobrychis viciifolia*) or on chicory (*Cichorium intybus*) had lower levels of this parasite compared to those receiving non CT-containing control feeds (Marley *et al.*, 2003b, Tzamaloukas *et al.*, 2005, Thamsborg *et al.*, 2004). With the same forages, however, essentially no effect was observed against the intestinal GIN species *Trichostrongylus colubriformis* (Tzamaloukas *et al.*, 2005, Athanasiadou *et al.*, 2005). It is unclear, whether these findings are evidence of true species specificity in the sense that CTs inhibit or suppress biological processes in *Teladorsagia* but not in *Trichostrongylus* or whether they are a result of the different location of these parasites in the gastrointestinal tract. Overall, the accumulated data suggest that one important direction in this field of research must be the investigation of individual GIN species responses towards individual CT-containing forages in order to learn more about the specific direct action of CTs (Hoste *et al.*, 2006).
In this experiment, we intended to compare the efficacy of four tanniferous forages with respect to faecal egg excretion and worm burden of two GIN species in lambs under identical experimental conditions in order to get further insights in parasite and forage specific effects. As a first experimental parasite, we chose *Haemonchus contortus*, an abomasal sheep parasite of global importance with which only few studies have been conducted so far (Paolini *et al.*, 2003b, Lange *et al.*, 2006, Min *et al.*, 2004, Heckendorn *et al.*, 2006). To the best of our knowledge, the administration of fresh CT-containing forages to small ruminants infected with *Haemonchus* has never been evaluated yet. As a second parasite *Cooperia curticei* was included in the study, an intestinal sheep GIN of regional importance in Europe (Rehbein *et al.*, 1996, 1998).
As tannin containing forages we chose birdsfoot trefoil (*Lotus corniculatus*), big trefoil (*Lotus pedunculatus*) and sainfoin (*Onobrychis viciifolia*). Furthermore, chicory (*Cichorium intybus*) was included in the study. It is known from former studies that these forages differ widely in their tannin content. By quantifying the CT-content of each forage and by recording the individual feed intake of every lamb, we aimed to establish a dosage-effect relationship of CTs against *Haemonchus* and *Cooperia*

with CTs from field grown plants. A further objective of the study was to investigate whether the antiparasitic effect of CTs is a direct result of the elimination of established GIN or if the effect is limited to a temporary reduction in parasite fecundity. A number of studies found that for *Trichostrongylus colubriformis* and *Haemonchus contortus*, the observed reductions in egg excretion disappeared when CT-administration was stopped (Min *et al.*, 2005, Lange *et al.*, 2006, Athanasiadou *et al.*, 2000b), suggesting that CTs temporally reduced the female worm fecundity. Other experiments including a variety of CT-containing plants and GIN, however, concluded that reductions in FEC were mainly associated with reductions of adult worms (Niezen *et al.*, 1995, Thamsborg *et al.*, 2004, Heckendorn *et al.*, 2006).

5.2. Animals, Materials and Methods

5.2.1. Forage cultivation
In early spring 2004 four 0.25 ha plots were sown as pure stands of chicory (*Cichorium intybus*, cv. Grasslands Puna), birdsfoot trefoil (*Lotus corniculatus*, cv. Odenwälder), sainfoin (*Onobrychis viciifolia*, cv. Visnovsky), big trefoil (*Lotus pedunculatus*, cv. Barsilvi) or with a ryegrass / lucerne mixture at FiBL, Frick, Switzerland. Sowing rates were adjusted for germination percentages in the seed samples and corresponded to 32 kg ha^{-1} of germinable seed for ryegrass / lucerne, 18 kg ha^{-1} for birdsfoot trefoil, 180 kg ha^{-1} for sainfoin, 11 kg ha^{-1} for big trefoil and 4 kg ha^{-1} for chicory. All plots except for big trefoil were cut in mid-May and the re-growths were used as experimental feeds in late June. Big trefoil at this stage had to be excluded from the study, because this species was outcompeted almost completely, resulting in swards that contained less than 1 % DM of this forage. At the start of the experiment, sainfoin and birdsfoot trefoil were at the 50 % flowering stage whilst chicory was still in a vegetative stage.

5.2.2. Animals
Twenty-four Swiss White Alpine x Swiss Black-Brown Mountain lambs were used in the study. They were penned indoors under conditions that minimized nematode infection. Lambs were given a multivalent vaccination against clostridial infections at approximately 5 and 10 weeks of age and were treated with levamisole (Endex® 8.75 %, 7.5 mg/kg body weight) to ensure trichostrongle-free conditions. After weaning, at an age of 3.5 – 4 months, the animals were accustomed to fresh fodder during a 4 week period prior to the start of the experiment. At the start of the feeding trial the lambs had a mean live weight of 22.4 ± 0.6 kg.

5.2.3. Parasite isolates and experimental infection
Infective larvae of *Haemonchus contortus* and *Cooperia curticei* were cultured from faeces of monospecifically infected donor lambs according to standard procedures. Parasite isolates were kindly provided by Merial, Germany. Twenty-seven days prior to the start of the feeding experiment all lambs were inoculated with a single dose of 7000 third stage larvae of *Haemonchus contortus* and 15000 third stage larvae of *Cooperia curticei*. The severity of the infection in all lambs was quantified 24 days post infection (p.i.) or 3 days prior to the start of the CT feeding experiment, respectively, by means of faecal egg counts (FEC).

5.2.4. Experimental Design
The CT-feeding experiment was conducted in a randomised complete block design. On day 24 p.i., the lambs were allocated to six blocks according their initial FEC (see above). Within each block, the four sheep were randomly assigned to the four experimental feeds: chicory, birdsfoot trefoil, sainfoin or the non-tanniferous control feed (i.e. a ryegrass / lucerne mixture). From day 27 p.i. animals were fed with their respective experimental feed for 17 consecutive days. After the CT-feeding period, lambs were united to one flock and subjected to group feeding with non-tanniferous control feed for another 11 days in order to test whether potential effects on FEC of the different feeds on lambs are reversible. Finally lambs were slaughtered to determine the adult worm burden in their intestines.

5.2.5. Forage administration, feed intake and live weight
The daily required portions of each of the 4 forages were harvested early every morning and stored at 4° C for later use. Sub-samples of each forage were taken immediately after harvest and analysed daily for dry weight. Depending on the forage dry weight results of the previous day, lambs were offered a daily total of between 6 – 9 kg fresh fodders, in order to achieve close to *ad libitum* feeding conditions. Fodder was offered 3 times a day (morning, afternoon and evening) and individually

weighed before administration. Equally, fodder spillage was measured 3 times a day. Thus, daily fodder intake was known for each individual sheep. In addition, live weights of animals were recorded every week.

5.2.6. Feed analysis

The botanical composition and CT-contents of the experimental feeds were analysed on day 3, 8 and 13 of the 17 day CT-feeding period. In order to describe the botanical composition of each feed, harvest samples were separated according to plant species, dried and the relative contribution of different functional plant groups to total yield were calculated. Corresponding samples were freeze dried and analysed for condensed tannins by the method described in Terrill *et al.* (1992). Nutritive values, such as net energy, protein content and in vitro-digestibility were determined from a bulk sample of daily collected and lyophilized sub-samples at the end of the experiment using standard procedures. Organic matter digestibility (OMD) of the experimental feeds was determined in vitro according to Tilley and Terry (1963).

5.2.7. Parasitological procedures and measures

During the individual CT-feeding period and the subsequent group feeding period, individual faecal samples were taken from the rectum every 2-4 days. Faecal samples were processed immediately for FEC (Schmidt, 1971) and dry matter (DM) content. The DM of the faeces was calculated from a 3 g sub-sample dried in a force-draught oven at 105°C for 16 h. Since different feeds can influence faecal dry matter, faecal egg counts were expressed as the number of eggs per gram of dried faeces (FECDM) as described in Heckendorn *et al.* (2006). During the CT-feeding period, pooled quantitative faecal cultures were prepared group wise using a 2 g sub-sample of fresh faeces from every lamb (M. Larsen, personal communication). Prior to cultivation, the faecal material was homogenized thoroughly in order to uniformly distribute the eggs and pre-culture FEC were performed. Per culture 12 g of faecal material was placed in a polystyrene container and incubated for 10 days at 20°C under conditions that maximised humidity. Post-culture, larvae were extracted for 24 h using a Baermann apparatus and transferred to Falcon tubes. After a 6 hour storage period at 4° C, excess liquid was removed by siphoning and the concentrated larvae were transferred to tissue culture flasks and stored at 6° C until processing. Total numbers of infective larvae were counted in 500 µl aliquots (10 times 50 µl) and the mean counts extrapolated to the total culture volume. FECDM specific for *Haemonchus* and *Cooperia* were calculated as described in Heckendorn *et al.* (2006). Blood samples were regularly taken during the experiment in order to monitor packed cell volume (PCV) as a parameter reflecting the severity of the parasite infection. A PCV level below 15 was defined as exclusion criterion.

As during CT-feeding the dry matter of daily feed intake (DMDI) and the percentage of in vitro digestibile organic matter (OMD) were known for each feed, the dry matter of total daily faecal output (TDFO; Mayes *et al.*, 1986) could be estimated for each individual sheep:

TDFO (g DM) = DMDI x (1-OMD)

Subsequently, an estimate of the total daily faecal egg output (TDFEO) was obtained by multiplying the total daily faecal output (TDFO) by the faecal egg counts per gram dried faeces (FECDM):

TDFEO = TDFO x FECDM

TDFEO specific for *Haemonchus* and *Cooperia* were calculated on the basis of L3 percentages of the respective species determined in the cultures as described in Heckendorn *et al.* (2006).

On day 28, immediately after slaughter, the abomasa and small intestines were ligated, opened and washed thoroughly in order to collect the luminal contents. Adult worm counts and sex identification were performed in a 10 % aliquot.

5.2.8. Statistical analysis

FECDM, TDFEO and worm burden were analyzed in the end of the experiment by means of generalized linear models (GLMs) under the assumption of negative binomial distributed residuals. For FECDM and TDFEO analogous models were fitted at the end of the CT-feeding period. For the analysis of aggregated data (e.g. worm burden), such GLMs have the advantage that they allow untransformed data to be analysed and that statistical tests of significance for the parameters of the models have a higher statistical power and a reduced risk of type I and type II errors compared to corresponding tests with log-transformed data and models that assume normally distributed residuals (Torgerson *et al.*, 2005, Wilson *et al.*, 1996). Fodder intake, faecal output and animal live weight were analyzed in 'normal' linear regression models. Both, GLMs and normal regression models contained parameters for the different *blocks* and parameters for the different *feeds*. As the design matrices of all models were dummy-coded and the control fodder set as a reference category, tests of significance for each of the three fodder parameters corresponded to testing whether there are differences in the response variable between any of the groups that received CT-containing forages and the control fed group (ryegrass / lucerne).

In order to establish a dose-response relationship, the relative reductions in TDFEO between the start and the end of the CT-feeding period were regressed as linear functions of the individual cumulative CT-intake. No statistical analyses were performed on the agronomical-, the feed analytical- and the PCV data as these measurements aimed only at describing the forages and at monitoring the health status of the animals, respectively. All data were analysed using STATA® 9.0 (StataCorp LP, 4905 Lakeway Drive, Texas 77845, USA) software.

5.3 Results

5.3.1. Botanical feed analyses

An overview of the botanical composition of the different feeds during the CT-feeding period is provided in Figure 5.1. The control fodder had a ryegrass : lucerne ratio of 5:4 and the harvested samples contained over 90 % DM of sown species. In the harvests of the chicory, birdsfoot trefoil and sainfoin stands, the relative contributions of these species to the harvest were 84, 68 and 61 % DM respectively, when averaged over the whole CT-feeding period. Especially for sainfoin it is important to note that the relative contribution of that species in the harvest was not constant but increased from 46 to 74 % within the 17 days CT-feeding period. In all stands, unsown, competing species were mainly herbs such as *Taraxacum officinale* or legumes such as *Trifolium repens*.

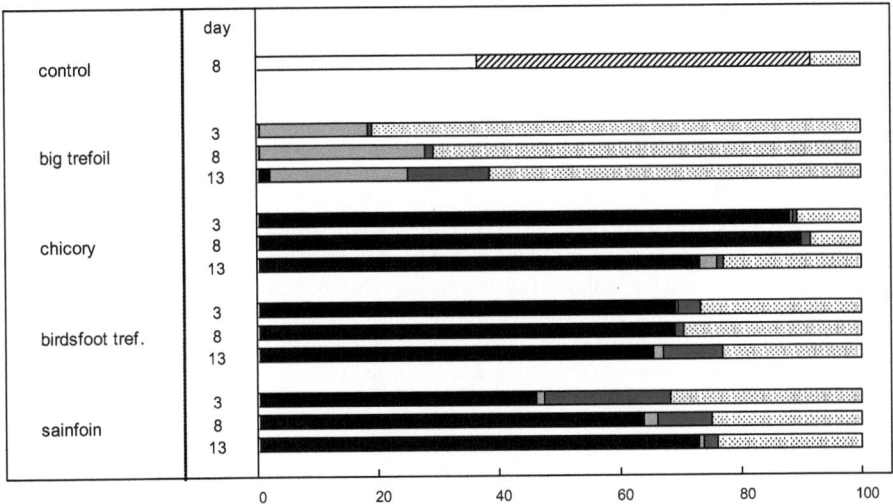

Figure 5.1. Botanical composition of the harvests during the CT-feeding period on days 3, 8 13 as functional groups: sown CT-containing forage (black), unsown legumes (light grey), lucerne (white), unsown grasses (dark grey), ryegrass (hatched) and unsown herbs (dotted).

5.3.2. Physical and chemical feed analyses

Physical and chemical key values of the different feeds are shown in Table 5.1. Dry matter contents of all CT-forages were relatively low and in vitro digestibility high compared to the control feed. Both differences were most pronounced for chicory where dry matter content was only half of that of the control forage but in vitro digestibility was 20 % higher. This will be important for the interpretation of faecal egg count data in the later text.

Averaged over the whole CT-feeding period, sainfoin (26 g CTs kg^{-1} DM) and birdsfoot trefoil (15 g CTs kg^{-1} DM) had higher contents of condensed tannins than chicory or the grass/lucerne mixture (both values < 5 g CTs kg^{-1} DM). In the case of sainfoin fodder, the CT-content was not stable but increased according to its increasing biomass proportion in the harvest from 19 g CTs kg^{-1} DM in the beginning to 34 g CTs kg^{-1} DM at the end of the CT-feeding period (Figure 5.1).

	Control	Chicory	Birdsfoot trefoil	Sainfoin
DM (g kg^{-1} fresh matter)	248	117	187	196
APD (g kg^{-1} DM)	85	92	98	93
NE (MJ kg^{-1} DM)	4.5	6.3	5.2	4.9
OMD (g kg^{-1} DM)	558	677	601	579
CT (g kg^{-1} DM)	0.2	3.1	15.2	26.1

Table 5.1. Mean dry matter (DM), absorbable protein at the duodenum (APD), net energy (NE), in vitro organic matter digestibility (OMD) and condensed tannin (CT) content of the experimental feeds.

5.3.3. Live weight, feed intake and faecal output

At the beginning of the CT feeding period, the average animal live weight was 22.4 ± 0.6 kg (mean ± SE). During the CT-feeding period, only animals of the sainfoin group consumed dry matter amounts comparable (930 ± 45 g day-1) to those consumed by the control fed animals (1050 ± 60 g day-1; Figure 5.2). Dry matter intake of lambs of the birdsfoot trefoil and the chicory groups was significantly lower compared to control fed lambs (27 % and 52 %, respectively, both P < 0.01). Daily live weight gains were generally low and in birdsfoot trefoil fed animals were comparable to those achieved by control fed animals (80 ± 30 g day-1 and 70 ± 40 g day-1, respectively). Highest daily weight gains where achieved in the sainfoin group (120 ± 20 g day-1) whereas for chicory fed animals no weight gain was recorded within the 17 days CT-feeding period. Although the differences in daily live weight gain between feeds were remarkable, this was not reflected in significant differences in live weight between the feeding groups at the end of the study (data not shown); probably because of compensatory fodder intake during the 11 days of control feeding. Due to the differences in dry matter intake and digestibility between the different diets, faecal output was significantly affected by the forage treatment (P < 0.001). Compared to the control animals (408 ± 25 g), the overall mean faecal output during the CT-feeding period was reduced by 293 ± 26 g and 142 ± 26 g faecal dry matter for chicory and birdsfoot trefoil (both P < 0.01) and by 56 ± 27 g (P < 0.05) in the sainfoin group.

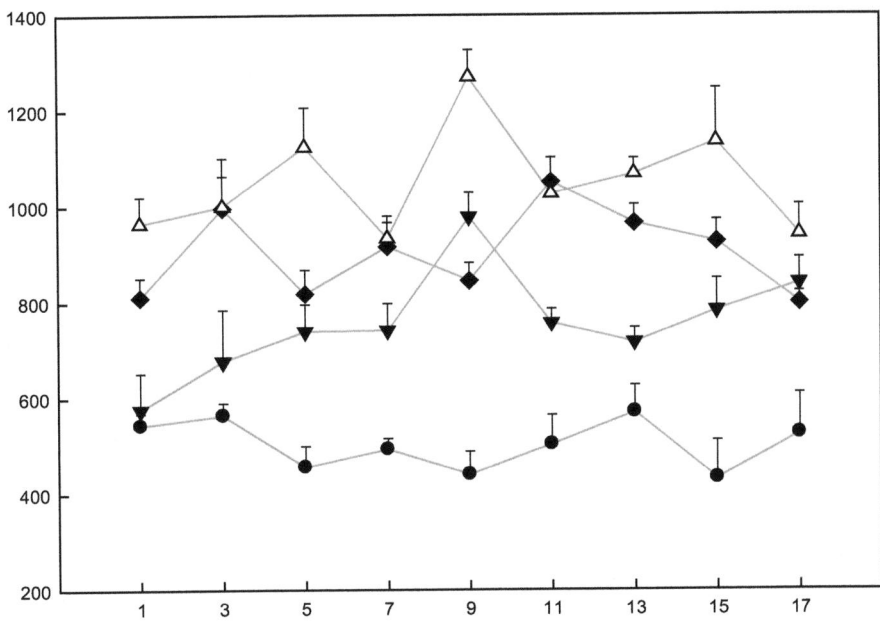

Figure 5.2. Mean daily dry matter intake of lambs consuming ryegrass/lucerne (open triangles), chicory (closed circles), birdsfoot trefoil (closed triangles) or sainfoin (closed diamonds) during the CT-feeding period. Bars indicate SE's of the means. The dotted line symbolises the expected DM intake of a lamb with mean live weight of 22 kg (equals the mean live weight of lambs included in the experiment) and assuming moderate live weight gain of 200 g day^{-1} as given in Arrigo et. al. (1994).

5.3.4. Faecal egg counts (FECDM) and daily egg output (TDFEO)

Regular recordings of FECDM during the entire experiment are presented for each individual lamb in Figure 5.3. Infection intensities were already highly variable at the beginning of the experiment. By allocating the animals to blocks according to their initial FEC (i.e. 24 days p.i.), it was possible to obtain four comparable groups and to control this nuisance parameter when analysing the effect of the different forages at the end of the study. Average FECDM of the control fed group were fairly stable at about 15'000 eggs per gram of dried faeces from the beginning of the feeding experiment until the animals were slaughtered. For the chicory fed-group, FECDM increased markedly during an early phase of the CT-feeding period, partly because of a concentration effect due to the reduced fodder throughput. One animal of this group had to be removed from the study after 11 days of experimental feeding (i.e. 38 days p.i.) because PCV fell below a value of 15 (exclusion criterion). After this initial increase, FECDM started to normalize in chicory fed animals five days after the start of the CT-feeding and at the end of the CT-feeding period, FECDM was non-significantly reduced by 44 % compared to control fed animals (Table 5.2). This reduction persisted in the subsequent 11 days period of control feeding. Birdsfoot trefoil and sainfoin feeding decreased FECDM instantly and rapidly, arriving at FECDM reductions of 47 % (P = 0.11) and 57 % (P < 0.05) in the end of the CT-feeding period compared to control, respectively (Figure 5.3, Table 5.2). For both feeds, FECDM remained low compared to the control after CT-feeding ceased. The final FECDM of birdsfoot trefoil or sainfoin fed animals, just before the slaughter of sheep, were reduced by 65 % (P < 0.01) and 32 % (P = 0.24) compared to controls, respectively (Table 5.2). All CT-containing forages used in this experiment significantly reduced the FECDM specific to *Haemonchus*, but none of them reduced the FECDM specific to *Cooperia* (Table 5.2).

Compared to the control group and consistent with FECDM, all CT-fed groups had significantly reduced daily faecal egg outputs (TDFEO) at the end of the CT-feeding period (all CT-fed groups: P < 0.01). TDFEO was reduced by 81, 53 and 58 % compared to controls for chicory, birdsfoot trefoil and sainfoin, respectively. Also in line with FECDM was the observation that the reduction of faecal egg output was solely due to a reduction of *Haemonchus* specific TDFEO while *Cooperia* was apparently unaffected.

5.3.5 Worm burden

The worm burden of *Haemonchus* was consistently but not significantly lowered in all CT-fed groups when compared to the control and this reduction was more pronounced for female than for male worms (Table 5.2). In line with FECDM and TDFEO, no reductions of the *Cooperia* worm burden were observed for any feeding group.

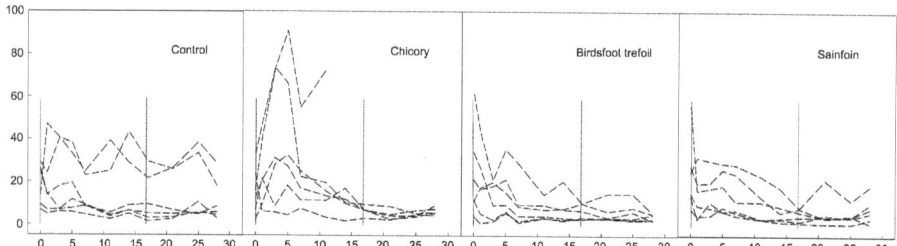

Figure 5.3. Faecal egg counts per gram dry faeces (FECDM) during the entire experimental period (N/group = 6). The dotted lines indicate the beginning and the end of the CT-feeding period. Summary statistics and statistical test results in the end of the CT-feeding period and in the end of the experiment are provided in table 5.2.
Note that in the chicory group one lamb had to be removed from the experiment on day 11.

Table 5.2. Mean faecal egg count per gram dry faeces (FECDM) and total daily faecal egg output (TDFEO) values at the end of the CT-feeding period and mean FECDM and mean worm counts at the end of the experiment of lambs in the control, chicory, birdsfoot trefoil and the sainfoin group. Percentages indicate the difference in means with respect to the control.

		Control	Chicory[1]	%	Birdsfoot trefoil	%	Sainfoin	%
End of CT feeding day 17	FECDM	11627	6534	−44 n.s.	6109	−47 n.s.	4993	−57 *
	Haemonchus	9418	2875	−69 **	3910	−58 *	3595	−62 *
	Cooperia	2209	3659	66 n.s.	2199	−1 n.s.	1398	−36 n.s.
	TDFEO (x105)	36.4	7.0	−81 **	17.1	−53 **	15.3	−58 **
	Haemonchus	29.9	3.1	−89 **	10.9	−63 *	11.0	−63 *
	Cooperia	6.9	3.9	−43 n.s.	6.1	−11 n.s.	4.3	−38 n.s.
End of experiment day 28	FECDM	11627	6367	−44 n.s.	3924	−65 **	7737	−32 n.s.
	Worm burden							
	Haemonchus ♀	285	168	−41 n.s.	120	−60 n.s.	148	−48 n.s.
	Haemonchus ♂	255	292	15 n.s.	155	−39 n.s.	202	−21 n.s.
	Cooperia ♀	1412	1530	8 n.s.	1775	26 n.s.	1645	17 n.s.
	Cooperia ♂	2215	2370	7 n.s.	2527	14 n.s.	2462	10 n.s.

Significance of statistical tests: n.s. not significant, * P < 0.05, ** P < 0.01
† Note that for chicory means and relative reductions of parasitological measurements were calculated without the lamb removed from the experiment at day 11 of the study

5.4. Discussion

5.4.1. Are differences of CT-action against GIN related to the host organ?

Our results suggest that all investigated tanniferous forage plants were active against *Haemonchus contortus* but none against *Cooperia curticei*. Concerning the controlled administration of fresh CT-containing fodder plants, this study for the first time demonstrated anthelmintic effects towards *Haemonchus*, one of the most important GINs worldwide. Compared to the control, faecal egg counts (FECDM) and total daily faecal egg outputs (TDFEOs) specific to *Haemonchus* were reduced consistently in all bioactive forage groups after 17 days of CT-feeding. These results were in accordance with tendencies of reduced *Haemonchus* worm recoveries at the end of the experiment (chicory 15 %, birdsfoot trefoil 49 %, sainfoin 35 %). Surprisingly and interestingly our data suggest that CTs affected the female worms more severely than the male worms (Table 5.2). As the mode of CT-action is unknown, it is difficult to plausibly explain this apparent sex bias. It is, however, a phenomenon that was observed consistently in all tested forages and also in a previous experiment (Heckendorn et al., 2006).

In contrast to *Haemonchus* and irrespective of the CT-containing fodder used, no reductions of adult *Cooperia* worm burdens were found. This result is in line with data obtained by Niezen *et al.* (1998a) who found that 42 days of birdsfoot trefoil feeding did not reduce the established *Cooperia* worm burden when compared to the control. The CT-concentrations used by Niezen *et al.* (1998a) were comparable (20-30 g CTs kg^{-1} DM) to the tannin concentrations used in our experiment (max. 26.1 g CTs kg^{-1} DM in sainfoin). However, experiments with higher concentrations of CTs suggested that *Cooperia* is not inherently resistant against condensed tannins. For example, faecal egg counts of *Cooperia* were significantly reduced when sainfoin silage or hay with CT-concentrations of 42 and 61 g CTs kg^{-1} DM, respectively, were fed for the same period of time as in the experiment presented here (Heckendorn *et al.*, 2006). Additionally, in the same study as mentioned above (Niezen *et al.*, 1998a), it was found that the *Cooperia* worm burden was significantly reduced by 37 % when sulla with a CT-content of 80-100 g CTs kg^{-1} DM was administered. Thus, the available data suggest that higher concentrations of condensed tannins are needed for the treatment of *Cooperia* than for *Haemonchus*. A comparison with recent results of in vivo experiments with tanniferous forages in sheep infected with *Teladorsagia circumcincta* and/or *Trichostrongylus colubriformis* suggests that also for this pair of GINs higher CT-levels are necessary to produce an antiparasitic effect on the intestinal species, whereas the abomasal species is affected already at lower CT-concentrations (Hoste *et al.*, 2006). Taken together, it could be speculated that the antiparasitic effects of tanniferous forages generally is achieved at lower CT-levels in the abomasum than in the small intestine and therefore would rather be organ dependent rather than GIN species related. Although this pattern has repeatedly been observed for field grown tanniferous forages, results obtained in a study with quebracho (a tannin-rich extract from the bark of the subtropical tree *Schinopsis spp.*) demonstrated that the abomasal nematode *Haemonchus* was unaffected by repeated drenches with high doses of quebracho (80 g CTs kg^{-1} DM) for 3 days (Athanasiadou *et al.*, 2001a). The apparent lack of effect on *Haemonchus* in the study by Athanasiadou *et al.* (2001a) could be a result of the short exposure of the worms to quebracho CTs (i.e. 3 days) compared to studies with forage CTs. Furthermore, discrepancies in anthelmintic effect due to the differences in the administered formulation (i.e. fodder CTs actively extracted by the host, quebracho CTs administered as a drench) cannot be excluded.

5.4.2. Plant specific anthelmintic activity of condensed tannins

Although all CT-containing feeds are highly palatable to sheep (Lüscher *et al.*, 2005), feed intake in the experiment was generally low. This was most probably related to the combined effect of the parasite infections and the high temperatures in summer 2004. Nevertheless, all tanniferous forage plant species investigated in this study showed antiparasitic effects with respect to at least some of the relevant parasitological measurements. However, the apparent anthelmintic activity of chicory (*Cichorium intybus*) seems puzzling considering that the CT-concentration of that fodder was very low (3 g CTs kg^{-1} DM). Yet, anthelmintic effects have repeatedly been demonstrated for chicory (Marley et al., 2003b, Thamsborg et al., 2004, Tzamaloukas et al., 2005). Possibly chicory owes its antiparasitic effect not only to CTs but also or alternatively, to the high content of phenolic metabolites other than CTs (Tzamaloukas et al., 2005) or to its content of sesquiterpene lactones for which anthelmintic effects have been shown in vitro (Molan et al., 2003a). Based on the results of our experiment, however, the feeding of young lambs with chicory cannot be recommended. We could not detect any live weight gain during the 17 days CT-feeding in the chicory group possibly due to a limitation of dry matter intake by the small rumen size and the high water content of this forage (see also section 5.4.5.). Furthermore, the chicory group was the only group from which an animal had to be excluded because of haemonchosis (PCV < 15).

Birdsfoot trefoil (*Lotus corniculatus*) reduced adult *Haemonchus* parasite burden, faecal egg counts and total daily faecal egg output consistently by about 50 % compared to the control fed animals. Other studies examining the anthelmintic effect of freshly administered birdsfoot trefoil exclusively worked with natural GIN infections and it is therefore difficult to compare these results with the findings of our study. The majority of these experiments found no effect of birdsfoot trefoil feeding on GIN (Hoskin et al., 2000, Niezen et al., 1998a, Bernes et al., 2000). Only in one experiment, feeding this fodder to growing lambs was associated with a reduced abomasal worm burden and reduced FEC, which is in accordance with our findings (Marley et al., 2003b).

Sainfoin (*Onobrychis viciifolia*) had the highest content of condensed tannins of all tested plants and showed similar activity against GIN as birdsfoot trefoil while allowing a higher daily weight gain than the one achieved by control fed animals. Antiparasitic effects of sainfoin condensed tannins (and also of flavonol glycosides) have been confirmed in vitro (Barrau et al., 2005) and in vivo (Thamsborg et al., 2004). Thamsborg et al. (2004) found 80 % FEC reduction and 35 % reduced adult *Teladorsagia* worm numbers in lambs co-infected with *Trichostrongylus vitrinus* after 3 weeks of sainfoin grazing when compared to a grass-clover fed control. Recent studies suggested that the anthelmintic effect of sainfoin is also preserved in sainfoin hay and silage (Paolini et al., 2003b, 2005a, Heckendorn et al., 2006). Preservation of sainfoin, particularly in the form of silage, seems very promising: It can be used for the control of GIN independent of the season and more of the forage plants' leaflets which are particularly rich in condensed tannins are retained than during the drying and conditioning involved in hay-making (Heckendorn et al., 2006, Häring et al., 2007).

5.4.3. Is there an in vivo dose-response relationship?

It is widely accepted that the antiparasitic effect associated with the feeding of sulla (*Hedysarum coronarium*), sainfoin, big trefoil and birdsfoot trefoil can to some extent be attributed to their content of condensed tannins. In part, this view is supported by in vivo studies with quebracho and by in vitro studies with various sources of tannin (Athanasiadou et al., 2000b, Molan et al., 2000a, 2000b, 2000c, 2003a). However, causality between the antiparasitic effect associated with the feeding of any of the above mentioned bioactive forages and the tannins within them has never been demonstrated in an in vivo experiment. Most in vivo experiments have been conducted by comparing the effect of a tanniferous forage with that of a non-tanniferous one based on a completely different species and therefore a different control forage in many respects. Confounding with and interference of other primary and secondary metabolites cannot have been ruled out therefore. The only experiment known to us that used the same forage – big trefoil (*Lotus pedunculatus*) - both as treatment and as control (but in the control polyethylene glycol (PEG) was used in an attempt to inactivate CTs) failed to demonstrate that condensed tannins are responsible for the anthelmintic activity (Niezen et al., 1998b). An in vivo dose response curve for any forage plant and any parasite species would help to substantiate the role of condensed tannins as antiparasitic compounds. As in this study detailed data on fodder intake and on the CT-content of each forage were collected during the CT-feeding period, we related the CT-dose defined by the cumulative CT-intake to the relative reduction in total daily egg output (TDFEO) within the 17 days CT-feeding period (Figure 5.4). Since we did not find any effect on *Cooperia*, data are shown for *Haemonchus* only. The values for chicory fed lambs were excluded from calculation and Figure 4, because the anthelmintic effect of this fodder is probably associated to other secondary plant metabolites (see discussion above). For the remaining lambs, there was a slight trend across forage species for an increased antiparasitic effect with higher CT-concentrations. Particularly conspicuous however, are the strong relative reductions in faecal egg output in four lambs that received non-CT containing control fodder (lower left corner of Figure 5.4). Interestingly, these were four sheep that had a comparatively low infection at the beginning of the CT-feeding period whereas the two strongly infected sheep in the control group still had strong infections at the end of the experiment. In contrast, the strongest relative reductions of faecal egg outputs in birdsfoot trefoil and sainfoin fed animals was due to a reduction of egg output in sheep with the strongest infection at the start of the CT-feeding period. Overall, evidence indicates that a relationship between the cumulative CT-dose and the relative reduction in faecal egg output for *Haemonchus* might exist. Further dosing trials specifically addressing the question of dose-response relationships are necessary in order to get further insights to dose effects of individual sources of CT towards GIN.

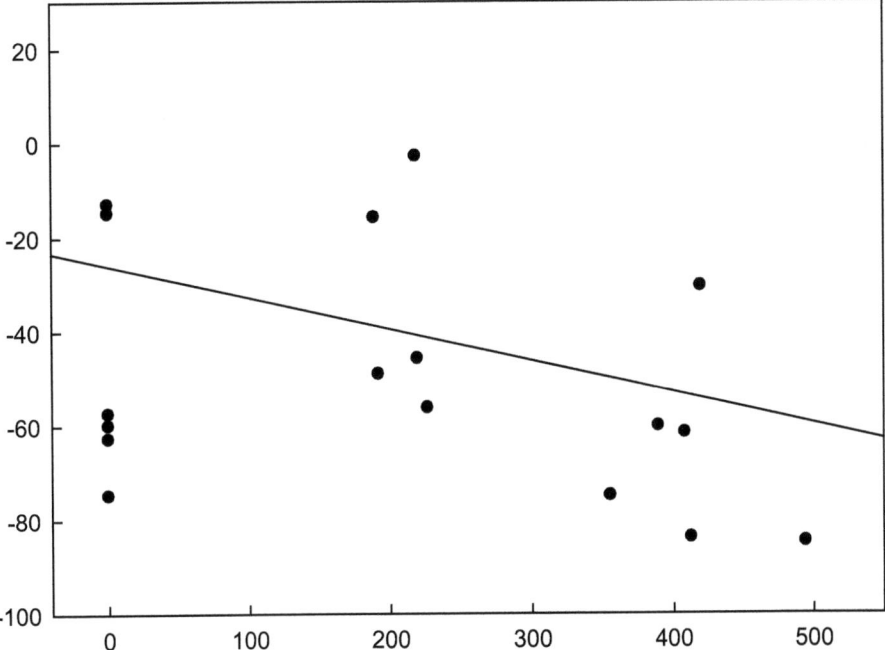

Figure 5.4. Relative total daily egg output (TDFEO) reduction (%) related to cumulative CT-intake during the 17 days CT-feeding period of control, birdsfoot trefoil and sainfoin fed animals. One value (X: 145, Y:133) is not shown in the graph but is included in the calculation of the regression line.

5.4.4 Reversibility of parasitological effect

The most commonly reported effect of tanniferous fodder plants is a substantially decreased faecal egg count (Hoste et al., 2006). Although this effect can be of great importance with respect to pasture contamination it is not necessarily a proof of a reduced worm burden or an improved health condition of the host. In fact, studies with goats found that reductions in FEC were essentially related to reductions in female worm fecundity with a potential of a renewed increase of FEC when the exposure to CT is withheld (Paolini et al., 2005a, . Similarly, Min et al. (2004) and Lange et al. (2006) found only temporarily reduced Haemonchus FEC after Sericea lespedeza (*Lespedeza cuneata*) hay feeding. In these studies FEC recovered to before treatment levels after the cessation of CT-feeding. In other experiments that were conducted mainly with sheep and relatively high concentrations of CTs, it has been observed that FEC reductions were a true consequence of a reduced worm burden (Scales et al., 1994, Niezen et al., 2002a, Thamsborg et al., 2004, Niezen et al., 1998b, Heckendorn et al., 2006, Niezen et al., 1995). This is in accordance with the results of the present study which was specifically designed to detect reversibility effects of FECDM. We found that predominantly the reduced female Haemonchus worm burdens in the CT-fed groups were responsible for the FECDM reductions.

In our experiment, it is unlikely that any indirect mechanism of CTs against *Haemonchus* such as an enhanced immune response (resistance) mediated by improved protein availability could have been responsible for the observed anthelmintic effects as protective immunity against *Haemonchus* is developed in lambs only after the age of 6 months (Urquhart et al., 1966b, 1966a) which was not the case for our animals. Hence, CT-feeding was associated with an animal health gain in terms of a lower worm burden due to a direct detrimental effect of CTs against adult *Haemonchus* (mainly against female worms) and also with a reduced pasture contamination because of the sustainable reduced egg output – at least for the period examined in this experiment.

5.4.5 Interpretation of faecal egg counts can be ambiguous in feeding trials

There were major differences with regard to fodder intake between the four feeding groups. Most strikingly, dry matter intake of chicory fed animals was only about half compared to control fed lambs. We believe that rumen size has limited the fresh weight intake of the young lambs. Subsequently, the high water content and the comparatively elevated dry matter digestibility of chicory reduced the faecal output in this group compared to the control. Averaged over the CT-feeding period, faecal output of chicory fed animals was 72 % lower than of control fed ones. Of course, this was not without consequence with regard to faecal egg counts. Under the assumption of an equal worm burden and an equal fecundity of female worms, FECDM would have been 3.6-fold in chicory fed lambs compared to the control. Thus, the initial increase in FECDM at the start of the CT-feeding period in the chicory group can fully be attributed to the reduced fodder throughput. The fact that FEC is sensitive to dry matter content and digestibility of the feeds and also to the actual feed intake, casts some doubts on the usefulness of this parameter in feeding trials. Whenever possible, total daily faecal egg output (TDFEO) should be preferred to FEC when the effect of different fodders is to be compared.

5.5 Conclusions

We found that all investigated tanniferous forage plants were active against *Haemonchus contortus* but none against *Cooperia curticei*. It is proposed that in the case of birdsfoot trefoil and sainfoin, anthelmintic effects were principally related to the action of CTs, whereas for chicory other secondary plant metabolites such as sesquiterpene lactones need to be considered. *Cooperia* does not seem inherently resistant to CTs but probably higher levels of these compounds are needed to produce anthelmintic effects. Together with the available data from other studies our data suggest that the anthelmintic effect of CT-containing fodder is to some extent dependent on the location of the parasites in the gastro-intestinal tract. The sustainable reduction of faecal egg counts as well as the lowered *Haemonchus* burden at the end of the experiment suggested that the anthelmintic activity of CTs has directly affected the adult parasites of this species. Limitations of standard faecal egg count as an indirect measure for the severity of a parasite infection in feeding trials was demonstrated. Whenever possible, total daily egg output should be preferred to FEC because it is insensitive to differences in fodder throughput between feeding groups. Overall, birdsfoot trefoil and in particular sainfoin seemed promising in contributing to an alternative, integrated control strategy against GINs not only by mitigating parasite related health disturbances of the host but also by a sustained reduction of pasture contamination by reduced egg output.

6. Effect of sainfoin *(Onobrychis viciifolia)* silage and hay on established populations of *Haemonchus contortus* and *Cooperia curticei* in lambs

Based on: Heckendorn, F., Häring, D.A., Maurer, V., Zinsstag, J., Langhans, W., Hertzberg, H. (published in: Vet. Parasitol. 2006, vol. 142, p. 293-300)

6.1 Introduction

Nematode infections of the gastrointestinal tract represent a major constraint in sheep husbandry, resulting in significant production losses (Brunsdon and Vlassoff, 1982, Coop *et al.*, 1985, Sykes, 1994, Parkins and Holmes, 1989). For almost 50 years the control of these parasites has relied almost entirely on the repeated use of anthelmintics (Williams, 1997). There are however several factors highlighting the need to develop alternative approaches in gastrointestinal nematode (GIN) control. These include widespread anthelmintic resistance within worm populations (Jackson and Coop, 2000) and the concern of consumers for drug residues in animal products (Waller, 1999).

One complementary approach to reduce the dependence on anthelmintics is the use of tanniferous plants to limit nematode infections. Controlled indoor and outdoor studies with sheep have shown that the consumption of tanniferous legume forages like sulla (*Hedysarium coronarium*), big trefoil (*Lotus pedunculatus*) or birdsfoot trefoil (*Lotus corniculatus*) were associated with negative effects on host parasitism (Kahn and Diaz-Hernandez, 1999, Min and Hart, 2002, Niezen *et al.*, 1995, 1998b, Marley *et al.*, 2003b). In parasitized goats, promising results have recently been obtained with sainfoin (*Onobrychis viciifolia*) hay (Paolini *et al.*, 2003b, 2005a, Hoste *et al.*, 2005). These reports for the first time documented that the anti-parasitic effects were preserved when using a tanniferous legume in conserved form. To our knowledge, however, no experimental work exists with ensiled tanniferous plant material, although this conservation procedure is often preferred by farmers in regions with moderately warm summer temperatures, which limit the hay production of several fodder plants. Furthermore the use of conserved tanniferous plants (hay and silage) against GIN has never been evaluated in sheep.

Given the extensive body of knowledge accumulated in the last two decades, it is surprising that only few reports have focused on *H. contortus* (Paolini *et al.*, 2003a, Athanasiadou *et al.*, 2001a, Paolini *et al.*, 2005b) although this parasite is probably the most important sheep nematode world-wide. In addition to *H. contortus*, *C. curticei* was included in the study as a widespread intestinal species with regional importance (Rehbein *et al.*, 1996, 1998).

The objectives of the current study were, to determine the effects of ensiled and dried sainfoin on established populations of *H. contortus* and *C. curticei* in lambs and to assess the consequences of these two treatments on animal productivity.

6.2. Materials and methods

6.2.1. Animals

Twenty-four Swiss White Alpine x Swiss Black-Brown Mountain lambs (10 females and 14 males) were used in the study. They were reared in a common pen under conditions that minimized nematode infection. The lambs were 2.5 – 3 months old and had a mean live weight of 33.1 ± 0.1 kg at the start of the trial. All animals were treated with levamisole (Endex® 8.75 %, 7.5 mg/kg body weight) to ensure helminth-free conditions.

6.2.2. Forage and feed constituents
Four different experimental feeds were used. Sainfoin hay and silage were produced in summer 2004 from sainfoin (*cv.* Visnovsky) monoculture swards located at the Swiss Federal Research Station ALP (Posieux, Canton of Fribourg, 660 m above sea level) and at the Research Institute of Organic Agriculture (FiBL, Canton of Aargau, 350 m above sea level). For hay production the fresh sainfoin plant material was artificially dried for 48 h at 30°C using a vented drying chamber. Silage units of approximately 35 kg were produced by pressing the cut sainfoin at approximately 35 % dry matter (DM) and enwrapped in commercial silage film. Ryegrass / clover hay and a maize-lucerne silage were used as control forages respectively. At the beginning of the experiment in early 2005, CT-concentrations of all feeds were measured according to the butanol-HCl method described in Terrill *et al.* (1992). Feed constituents were determined as described in Arrigo (1994).

6.2.3. Parasite isolates
Infective larvae of *H. contortus* and *C. curticei* were cultured from monospecifically infected donor lambs according to standard procedures. Parasite isolates were kindly provided by Merial, Germany.

6.2.4. Experimental design and measurements
Twenty-eight days prior to the start of the feeding experiment all lambs were inoculated with a single dose of 7,000 third stage larvae of *H. contortus* and 15,000 third stage larvae of *C. curticei*. On the basis of faecal egg counts, individual weight and sex on day 24 post infection (p.i.), lambs were assigned to one of four experimental groups A-D consisting of 6 animals each. Groups A and B consisted of 2 female and 4 male animals each and starting from day 28 p.i. received sainfoin or ryegrass hay respectively for 16 consecutive days. Animals of group C and D consisted of 3 male and 3 female animals each and were fed with either sainfoin silage or with maize-lucerne silage for the same period. Lambs were offered the different feed *ad libitum*. On the basis of refusals per group, nutrient contents of the feeds and live weight of the lambs, the feeds were daily adjusted with a commercial concentrate (UFA 763; UFA AG, CH-6210 Sursee) or soy meal, in order to make them isonitrogenous and isoenergetic.

During the CT-feeding period individual faecal samples were taken from the rectum every 3-4 days for faecal egg counts and faecal cultures were made weekly for each feeding group. Faecal dry matter content was determined in a 3 g sub-sample dried in a force draught oven at 105 °C for 16 h immediately after collection. Since from day 3 onwards the faecal dry matter in the sainfoin silage group (pooled means ± S.E.M. 31 ± 1.5 %; $P < 0.05$) was significantly elevated compared to all other feeding groups, faecal egg counts were expressed as the number of eggs per gram of dried faeces (FECDM). Every week individual live weights were recorded and blood samples were taken for packed cell volume (PCV) measurements. At day 45 post infection all animals were slaughtered. Immediately after death the abomasa and small intestines were separated, opened and washed in order to collect the luminal contents. Adult worm counts and sex identification were performed in a 10 % aliquot.

6.2.5 Faecal samples and culture processing
Faecal samples were processed immediately for FEC (Schmidt, 1971). Pooled quantitative faecal cultures were prepared group-wise using a 2 g sub-sample of fresh faeces from every lamb (M. Larsen, personal communication). Prior to cultivation the faecal material was homogenized thoroughly in order to uniformly distribute the eggs and pre-culture FEC were performed. Per culture 12 g of faecal material was placed in a polystyrene container and incubated for 10 days at 20°C under conditions that maximised humidity. Post-culture, larvae were extracted for 24 h using a Baermann apparatus and transferred to Falcon tubes. After a 6 hour storage period at 4°C excess liquid was removed by siphoning and the concentrated larvae were transferred to tissue culture flasks and stored at 6°C until processing.

Total numbers of infective larvae were counted in 500 μl aliquots (10 times 50 μl) and the mean counts extrapolated to the total culture volume. Furthermore a total of 100 L3 larvae were differentiated within every culture. FECDM specific for *H. contortus* and *C. curticei* were calculated on the basis of L3 percentages of the respective species determined in the cultures, assuming equal development of the two GIN species (F. Borgsteede, personal communication). Per capita fecundity (PCF) was calculated separately for *H. contortus* and *C. curticei* by dividing the species specific FECDM recorded at slaughter by the total numbers of female worms recovered.

6.2.6 Statistical analysis

All data were analysed using STATA® 9.0 (StataCorp LP, 4905 Lakeway Drive, Texas 77845, USA) software. Evidence of aggregated distributions for both FECDM and worm burden was confirmed. Aggregated data are defined as the variance being greater than the mean (Torgerson *et al.*, 2005). For FECDM and worm burdens cross-sectional negative binomial regression models were therefore fitted separately for each point in time with the two parameters of the model being the arithmetic mean and the negative binomial constant. The mean egg count or worm burden and the 95 % negative binomial confidence intervals were estimated by maximum likelihood techniques. Comparisons were made between the (i) sainfoin hay and the control hay group and (ii) the sainfoin silage and the control silage group. Equivalent comparisons were done for the worm burdens. Per capita fecundity, PCV and live weight were analysed using t-tests.

6.3 Results

6.3.1 Nutritional contents and condensed tannin concentrations

Net energy contents of the feeds were comparable within the hay groups (sainfoin 5.1 MJ kg^{-1} DM, control 5.0 MJ kg^{-1} DM) and the silage groups (5.9 MJ kg^{-1} in both groups). Compared to control hay, sainfoin hay had a higher protein content (77 vs. 93 g/kg DM), whereas the two silage groups where essentially similar (70 g kg^{-1} DM both). Sainfoin hay had a higher CT-content than sainfoin silage (mean ± S.E.M, 6.12 ± 0.48 % DM and 4.19 ± 0.87 % DM). The CT-concentrations measured in the two control feeds were very low (mean ± S.E.M, hay 0.13 ± 0.01 % DM, silage 0.07 ± 0.03 % DM)

6.3.2. Consumption of feeds and live weight gain

There were no significant live weight differences between groups at the beginning of the study. All feeds were readily eaten by the lambs throughout the study period. Mean daily DM intakes per animal averaged over the entire experimental period were similar for all groups (approx. 1.2 kg DM d^{-1}, Table 6.1). No significant differences in daily weight gain were found between animals of the sainfoin silage group compared to those of the control silage group (mean ± S.E.M. 64 ± 27 g and 84 ± 20 g). There was a trend of increased daily weight gain in the sainfoin hay group compared to the control hay group (mean ± S.E.M. 163 ± 20 g and 96 ± 27 g; $P = 0.07$). However, no significant difference in live weight between the hay groups was present at the end of the study (sainfoin 35.8 ± 0.3 kg, control 34.8 ± 0.4 kg, P = 0.49).

			DM (kg) ± S.E.M	APD (g) ± S.E.M	NE (MJ) ± S.E.M
Hay fed	Group A	Sainfoin hay	0.98 ± 0.02	88.29 ± 1.23	5.34 ± 0.18
		Concentrate UFA 763	0.20 ± 0.01	22.76 ± 1.39	1.71 ± 1.39
		Total	**1.18**	**111.05**	**7.05**
	Group B	Control hay	1.27 ± 0.01	98.94 ± 0.73	6.46 ± 0.05
		Concentrate UFA 763	0.06 ± 0.01	6.35 ± 0.53	0.51 ± 0.04
		Total	**1.33**	**105.29**	**6.97**
Silage fed	Group C	Sainfoin silage	1.15 ± 0.01	72.29 ± 2.04	6.43 ± 0.06
		Soy meal	0.06 ± 0.01	18.03 ± 1.05	0.48 ± 0.04
		Total	1.21	90.32	6.91
	Group D	Control silage	1.09 ± 0.02	77.12 ± 1.07	6.37 ± 0.06
		Soy meal	0.15 ± 0.03	17.18 ± 1.34	0.46 ± 0.05
		Total	**1.24**	**94.30**	**6.83**

Table 6.1. Mean daily intake of dry matter (DM), absorbable protein at the duodenum (APD) and net energy (NE) per animal of groups A-D averaged over the 16 days study period.

6.3.3. Faecal egg counts

For *H. contortus*, the reduction of specific FECDM associated with the feeding of sainfoin was substantiated in the course of the study (Figure 6.1). After 16 days of consecutive experimental feeding *H. contortus* specific FECDM was reduced by 58 % (P < 0.01) in the hay group and by 48 % (P = 0.075) in the silage group compared to the respective controls. For *C. curticei*, already three days after experimental feeding had started, lambs fed with sainfoin hay or silage had a significantly reduced specific FECDM (both tests P < 0.001, Figure 6.2) compared to the controls. These differences remained stable until the end of the study (81 % reduction in sainfoin hay group [p < 0.05], 74 % reduction in sainfoin silage group [p < 0.01]).

6.3.4. Worm burden and per capita fecundity

The total *H. contortus* burden was lowered by approximately 50 % by both sainfoin feeds and in the hay group this reduction was significant ($P < 0.05$; Table 6.2). The per capita fecundity of *H. contortus*, was not significantly different between both hay (sainfoin 31.3 and control 26.5 eggs g^{-1} female^{-1}, P = 0.60) and silage groups (sainfoin 18.7 and control 27.3, eggs g^{-1} female^{-1} P = 0.51). Total *C. curticei* worm counts were not substantially reduced by both experimental feeds compared to the control feeds (sainfoin hay 9 %, P = 0.58 and silage 14 %, P = 0.14, respectively). However, for *C. curticei* a significantly lower per capita fecundity was found between the sainfoin hay and the control hay group (0.46 and 2.28 eggs g^{-1} female^{-1}, $P < 0.001$) and also between the sainfoin silage and the control silage group (0.48 and 1.68 eggs g^{-1} female^{-1}; $P < 0.05$).

6.3.5. Packed cell volume

Until the end of the experiment there was no significant difference in PCV between the sainfoin silage group and the respective control group. PCV levels of the sainfoin hay group were significantly lower than in the control hay group at day 16 after onset of experimental feeding (mean ± S.E.M 31.2 ± 0.7 % and 33.2 ± 0.7 % ; $P < 0.05$) but were still in the physiological range (30 – 38 %).

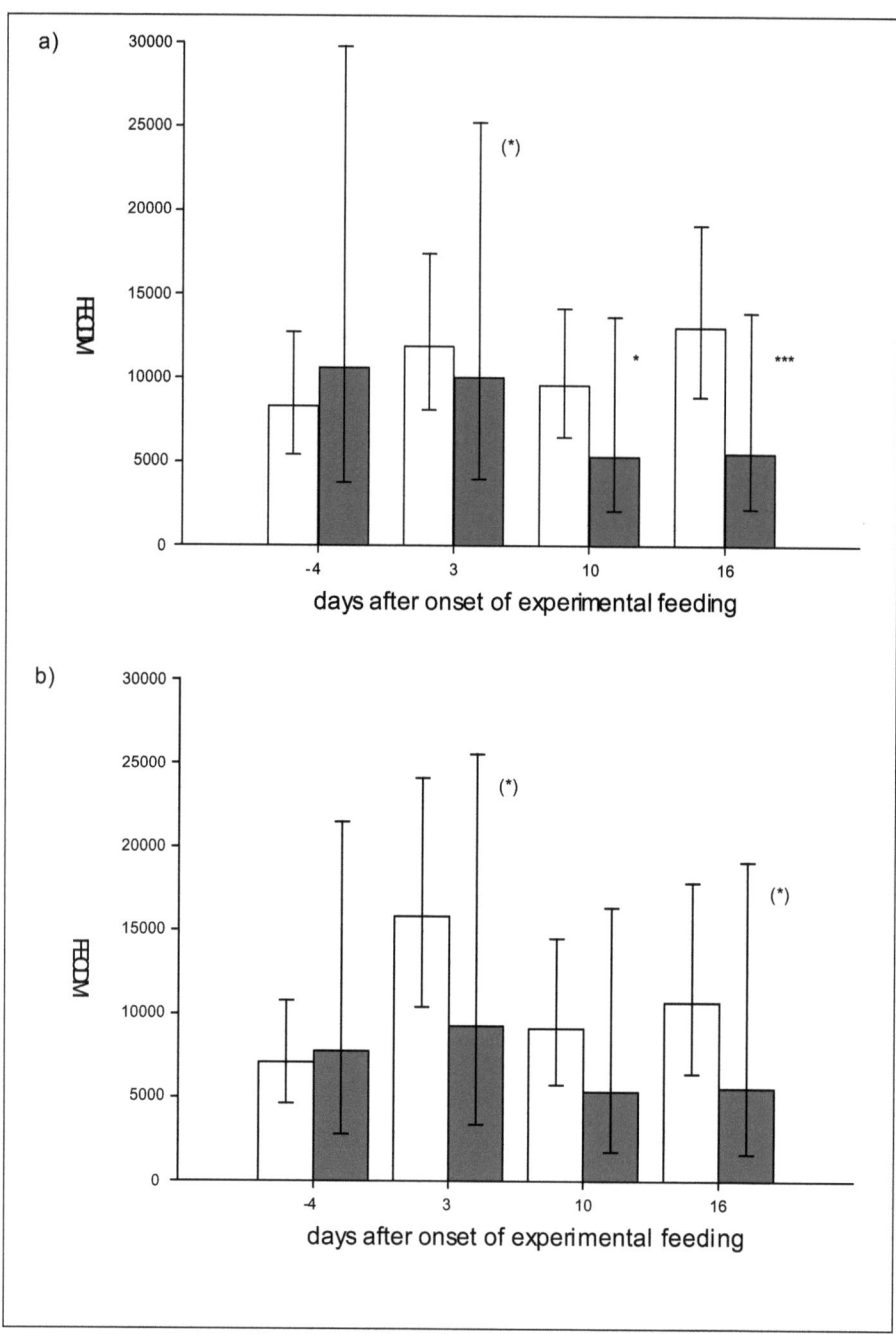

Fig 6.1. Comparison of FEC per gram faecal dry matter (FECDM) specific to *H. contortus* in group receiving (closed bars) or not (open bars) sainfoin hay (a) or silage (b). Error bars indicate 95% CI (Maximum Likelihood). Significance of statistical tests (*) P < 0.1 * P < 0.05   *** P < 0.001

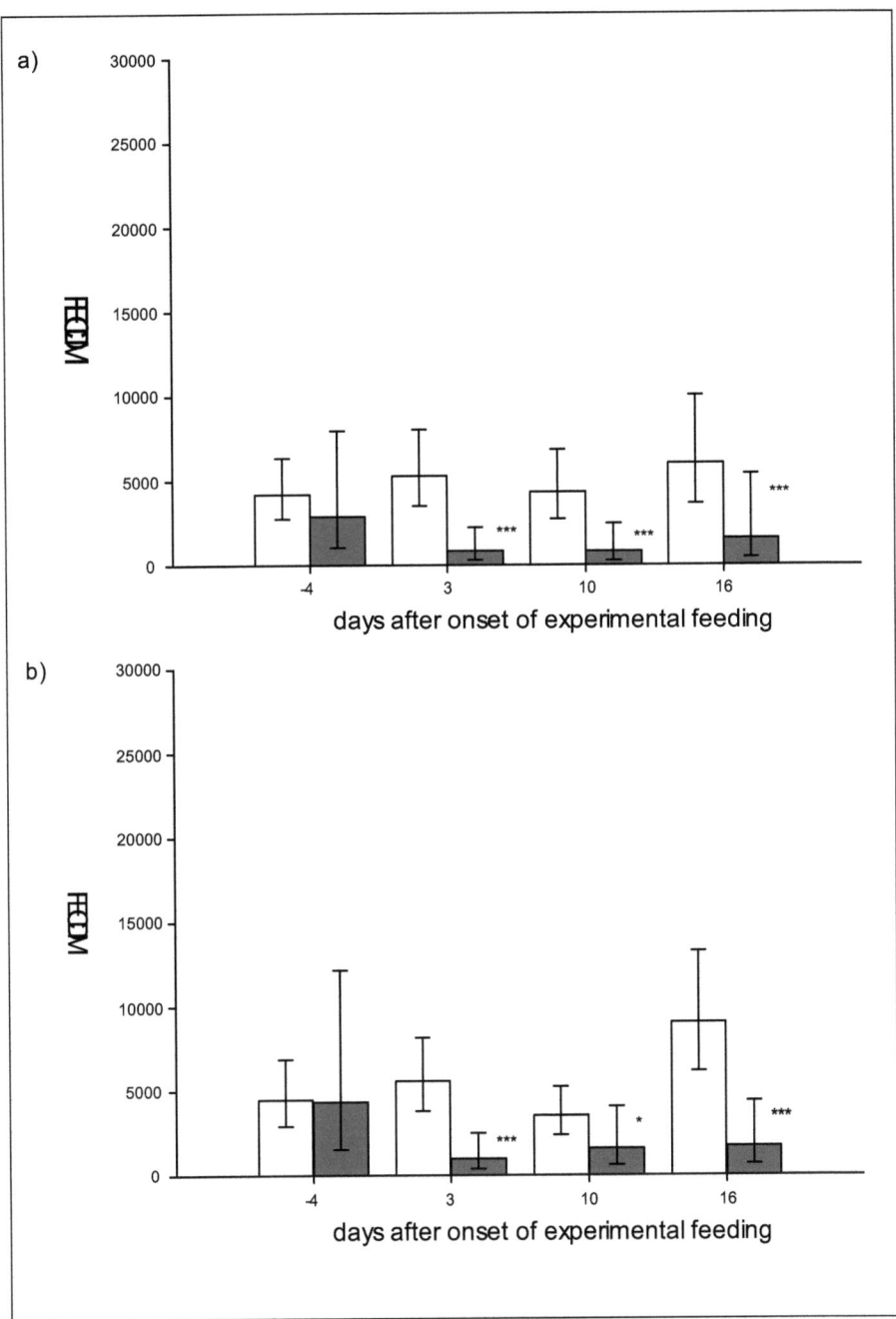

Fig 6.2. Comparison of FEC per gram faecal dry matter (FECDM) specific to *C. curticei* in group receiving (closed bars) or not (open bars) sainfoin hay (a) or silage (b). Error bars indicate 95 % CI (Maximum Likelihood). Significance of statistical tests ** $P < 0.05$ *** $P < 0.001$

Table 6.2. Means and (95 % CI maximum Likelihood) of *Cooperia curticei* and *Haemonchus contortus* worm counts. Percentages indicate the mean worm reduction of the sainfoin groups compared to the respective control groups. Significance of statistical tests: *n.s. not significant* $* P < 0.05$

	Haemonchus contortus						*Cooperia curticei*					
	♀	%	♂	%	Total	%	♀	%	♂	%	Total	%
Silage												
Control	542 (275 – 1067)		477 (268 – 847)		1018 (565 – 1835)		3742 (3352 – 4177)		2890 (2393 – 3490)		6632 (5772 – 7620)	
Sainfoin	257 (50 – 1319) n.s.	53	262 (65 – 1048) n.s.	45	518 (125 – 2150) n.s.	49	3142 (2409 – 4098) *	16	2582 (1637 – 4071) n.s	11	5723 (4092 – 8004) n.s	14
Hay												
Control	473 (312 – 719)		438 (279 – 688)		912 (594 – 1399)		3972 (3145 – 5015)		3183 (2464 – 2998)		7110 (5619 – 8997)	
Sainfoin	233 (85 – 807) *	51	250 (84 -743) n.s.	43	483 (172 – 1359) *	47	3695 (2104 – 6489) n.s.	7	2777 (1548 – 4981) n.s	13	6472 (3666 – 11423) n.s	9

6.4 Discussion

The main finding of this experiment is that by feeding sainfoin hay and silage for 16 consecutive days, the *H. contortus* worm burden was reduced by approximately 50 % compared to the corresponding controls. This is in contrast to other studies using condensed tannins, only documenting a decrease in FEC and fecundity of *H. contortus*. (Athanasiadou *et al.*, 2001a) administered different doses of quebracho CTs to *H. contortus* infected lambs and could not detect any difference in worm burden between drenched groups and control groups. Qebracho drenches were given only for 3 consecutive days, however, and the total study period was shorter than in our experiment (9 vs. 16 days). Paolini *et al.* (2003a) tested quebracho CTs given for 8 consecutive days at a concentration equivalent to 5 % of the dietary DM in goats artificially infected with *H. contortus*. No significant difference in worm burden was found after a total study period of 15 days when compared to a control group. It is known that the chemical composition of CTs may alter its bioactive potential (Aerts *et al.*, 1999). The results obtained with quebracho are therefore only of limited comparability to our study and must be interpreted carefully. A recent study evaluated the effect of repeated distribution of sainfoin hay in goats naturally infected with GIN, where the animals were fed with the hay for 7 consecutive days every month for a total period of 3 months (Paolini *et al.*, 2005a). At the end of the study only fecundity of *H. contortus* was decreased, but no difference was seen in adult worm burden compared to a control group. The prolonged administration period of 16 consecutive days in our study might explain the increased effect on adult *H. contortus* worms. It is theoretically possible that a longer exposure of worms to CTs is necessary in order to observe a nematocidal effect. This hypothesis is partly supported by a report focussing on another abomasal nematode (*Teladorsagia circumcincta*), where the adult worm burden was significantly reduced by 90 % when feeding sulla (another tanniferous legume) for a period of 42 days (Niezen *et al.*, 1994).

In our work no differences in per capita fecundity of *H. contortus* females were observed between the sainfoin and the control groups. This finding is in line with results of Athanasiadou *et al.* (2001a), where per capita fecundity of this parasite was unaltered in a quebracho fed group of lambs compared to a control group. In a study with goats it was found that the fecundity of *H. contortus* was significantly decreased when feeding sainfoin hay (Paolini *et al.*, 2005a). In that study, however, fecundity was assessed directly by counting eggs in utero and the observed difference to our results might be related to the difference in the applied methodology.

Compared to results obtained with *H. contortus*, we found only moderate reductions in adult *C. curticei* worm burden regarding both sainfoin feeds when compared to the respective control groups. This finding is in contrast to results obtained by Niezen *et al.* (1998a) who reported a significant lower *C. curticei* worm burden in lambs feeding on sulla for 42 days compared to a control group. Studies looking at the effect of CT on other intestinal nematode species reported contradictory results. Significant decreases in adult *Trichostrongylus colubriformis* burden have repeatedly been observed (Paolini *et al.*, 2005a, Athanasiadou *et al.*, 2000c, Niezen *et al.*, 1995), but also negative results have been reported (Niezen *et al.*, 1998b). Unfortunately these studies are of limited comparability because unequal experimental designs and different sources of CTs were used. Concerning per capita fecundity however, data from different experiments on intestinal nematodes seem to be much more coherent. In our experiment the per capita fecundity of *C. curticei* was significantly lowered by both sainfoin feeds. This finding is in accordance with studies on *T. colubriformis* where per capita fecundity was lower when quebracho CTs were administered (Athanasiadou *et al.*, 2000a, 2001a).

The results of the present study suggest that the significant decrease in FECDM is driven by two different mechanisms for the investigated nematode species. For *H. contortus* the decrease in FECDM seems to be mainly associated with the reduced worm burden, whereas for *C. curticei* the lower fecundity is suggested to be the relevant parameter in lowering FEC. Overall, the observation of decreased FEC is in agreement with work done by (Paolini *et al.*, 2003b, 2005a), where FEC was significantly reduced when sainfoin hay was administered to goats naturally infected with GIN.

In our experiment the *C. curticei* FECDM in the sainfoin silage group was already 31 % (P = 0.223) lower compared to the control hay group at the beginning of the experiment. This initial imbalance arose out of the delay of culture results and the need of a stratified group allocation according to FEC, which we were forced to do on total FEC. Still, a strong decline of the *C. curticei* specific FECDM was observable in the sainfoin silage group when compared to the control.

It is well documented that an increase in feed protein will improve the resistance of sheep to GIN (Coop and Holmes, 1996, Van Houtert and Sykes, 1996). Resistance effects mediated by an improved immune response were not expected within the chosen experimental period (i.e. 16 days) and due to the age of the animals used (Urquhart *et al.*, 1966b, 1966a). Nevertheless, feeds in our work were adjusted for protein and energy in order not to disguise possible CT-effects. With respect to PCV levels, a significant difference was only observed between the sainfoin hay and the control hay group. However, this difference was not of physiological relevance. A likely explanation for the overall observed equality of PCV levels between sainfoin and control groups is a delay in response of the parameter to *H. contortus* worm burden with respect to the short experimental period. Work done by Paolini *et al.* (2005a) showed a significant reduction of PCV levels in control hay fed goats compared to sainfoin hay fed ones only 70 days post infection, thus highlighting the need of a prolonged experimental period in order to observe any effect on this pathophysiological parameter.

Concerning preserved sainfoin, this study for the first time presents results pointing to a nematocidal effect towards *H. contortus*. The physiological basis of the underlying interactions is still unclear and remains to be elucidated. In regions with moderate climatic conditions the production of soil dry sainfoin hay is problematic because the cut plant needs a short and hot drying phase in order not to lose the CT-containing leaves in the drying process. As an easily feasible conservation alternative, ensiled sainfoin was therefore produced for this experiment. Although CT-contents were slightly lower in sainfoin silage as in hay, the anti-parasitic effect was also present when using this conservation method. Further studies using sainfoin silage must be performed, in order to determine its effect on other GIN species and to evaluate the acceptance of the strategy among farmers.

Overall, conservation of tanniferous fodder plants offers exciting opportunities with respect to centralized production, sale, storage and an extended administration independent of the season. However, its potential broader use should also be subjected to an analysis of its profitability.

7. On farm administration of sainfoin *(Onobrychis viciifolia)* silage to ewes naturally infected with gastrointestinal nematodes: effect on periparturient egg rise

Based on: Heckendorn, F., Maurer, V., Senn, M., Hertzberg, H., 2007 (in prep.)

7.1. Introduction

The epidemiology of gastrointestinal nematode (GIN) infections of sheep is strongly influenced by a decrease in immunological response of ewes around parturition (Michel, 1976, Barger, 1993). Beside increased susceptibility to newly acquired infections, this phenomenon is also characterized by a higher rate of development of arrested larvae to adults (Michel, 1974). Clinical disease may occur in ewes during this period, and the substantial increase of parasite egg excretion may represent a major source of infection for young immunologically naïve lambs. Traditionally, strategic treatment around parturition with an anthelmintic was used to prevent clinical disease in ewes and also to diminish the risk of GIN infection of lambs (Williams, 1997). In view of the widespread resistance of worm populations against conventional anthelmintics (Jackson and Coop, 2000) and the increasing concerns of consumers about drug residues in animal products, alternative GIN control strategies have been investigated in the last decade (Waller and Thamsborg, 2004).

The accumulated data from controlled indoor and outdoor studies suggest that the consumption of tanniferous legume forages may reduce the level of parasitism in sheep and thus reduce the reliance on anthelmintics (Hoste *et al.*, 2006). Amongst those, sainfoin (*Onobrychis viciifolia*) has repeatedly been shown to reduce GIN egg output and adult worm populations in goats and sheep and also to impair the establishment of new infections with some GIN species (Paolini *et al.*, 2003b, 2005a, Hoste *et al.*, 2005, Thamsborg *et al.*, 2004). More recently it was found that the conservation of sainfoin as silage did not reduce the antiparasitic effect (Chapter 6) and therefore allows the administration independent of the season. Furthermore, sainfoin is considered a good quality forage with high nutritive values (Borreani *et al.*, 2003).

Given the important epidemiological consequences of the periparturient egg rise (PPR) in ewes, it is surprising that the use of tanniferous legume forages has never been evaluated as an alternative control strategy during this period. The objective of the present study was therefore, to investigate the potential anthelmintic effect of sainfoin silage as measured by faecal egg count during the relaxation of immunity in periparturient ewes under on-farm conditions.

7.2. Materials and methods

7.2.1. Animals

Thirty-three, 3-5 year old Swiss white alpine ewes where used for the study. The animals were commonly grazed during the entire vegetation period 2005 on pastures pre-contaminated with GIN and being situated at an altitude of 400 – 500 m above sea level. For lambing in the winter period (February - April 2006), the ewes were housed in a common pen. No anthelmintic treatment was given at any time. Approximately 60 % of the experimental animals had lambed shortly before the start of the feeding experiment and 40 % where expected to lamb within 2 weeks.

7.2.2. Forages

Silage was produced in summer 2005 from sainfoin (*cv.* Visnovsky) monoculture swards located at the Research Institute of Organic Agriculture (FiBL, Canton of Aargau, 350 m above sea level). Silage bales of approximately 35 kg were produced by pressing the cut forage at approximately 35 % dry matter (DM) and enwrapping it in commercial silage film. A ryegrass / clover silage was used as control forage. CT-concentrations of the feeds were measured over the entire experimental period according to the butanol-HCl method described in Terrill et al. (1992).

7.2.3. Experimental design and measurements

Five days prior to the start of the feeding experiment faecal samples of 60 ewes were collected and parasite egg counts were determined as described in Schmidt (1971). All animals with FEC > 300 eggs per gram fresh faeces were selected for the experiment and assigned to either the sainfoin or the control group on the basis of FEC levels and the parturient status in order to produce comparable groups. The sainfoin and the control group consisted of 16 and 17 animals, respectively. From day zero of the experiment, the animals of the two groups were kept in separate pens and were *ad libitum* group-fed exclusively with the respective forage for 25 days. During the experimental feeding period individual faecal samples were taken from the rectum every 3-4 days for the determination of parasite eggs. Faecal dry matter content was individually determined for every animal in a 3 g sub-sample dried in a force draught oven at 105 °C for 16 h immediately after collection. Since from day 6 after the start of experimental feeding onwards the faecal dry matter in the sainfoin silage group (pooled means ± S.E.M. 31 ± 1.5 %; $P < 0.05$) was significantly elevated compared to the control feed, FEC were expressed as the number of eggs per gram of dried faeces (FECDM) as described in Heckendorn et al. (2006). Pooled larval cultures were prepared group-wise using a 2 g sub-sample of fresh faeces from every lamb at the beginning and at the end of the 25 days feeding period. After 10 days of incubation at 20 °C, larvae were extracted for 24 h using a Baermann apparatus and differentiated as described in MAFF (1986).

7.2.4. Statistical analysis

All data were analysed using STATA® 9.0 (StataCorp LP, 4905 Lakeway Drive, Texas 77845, USA) software. Evidence of aggregated distributions for FECDM was confirmed. Aggregated data are defined as the variance being greater than the mean (Torgerson et al., 2005). For FECDM cross-sectional negative binomial regression models were therefore fitted separately for each point in time with the two parameters of the model being the arithmetic mean and the negative binomial constant. The mean egg count and the 95 % negative binomial confidence intervals were estimated by maximum likelihood techniques. Comparisons were made between the sainfoin silage and the control silage group. FECDM AUC (Area under the Curve) data where analysed by means of a t-test. Data where transformed prior to analysis (log (x+1)) in order to stabilise the variances.

7.3. Results

7.3.1. Consumption of feeds and condensed tannin concentrations

After a few days, both experimental feeds were readily eaten by the lambs throughout the study period with an average consumption of 1.8 kg DM day^{-1} head^{-1}. CT-contents of the control fodder where low (< 2 g kg^{-1}) throughout the study. The CT-concentrations of sainfoin silage bales used in the course of the study were variable ranging from 15-28 g kg^{-1} DM (mean 20 g kg^{-1} DM).

7.3.2. Faecal egg count and AUC

No differences in FECDM where present between the sainfoin and the control group at the start of the feeding experiment (Figure 7.1). After 10 days of consecutive sainfoin silage feeding FECDM was reduced by 60 % ($P < 0.01$) when compared to the control fed animals. FECDM of the sainfoin fed group consistently remained lower until the end of the study compared to the control (Figure 1). The area under the curve based on FECDM, calculated for the entire experimental period, was lowered by 32 % in the sainfoin group, but this reduction was not significant (P = 0.17; data not shown). When the AUC was calculated for the period after FEC reduction took place (i.e. day 10-24) only, the reduction was 65 % (P = 0.09).

7.3.3. Larval cultures

The main GIN genera present in the experimental animals as detected by third stage larvae differentiation are presented in Table 1. There were differences in GIN composition between the sainfoin and the control group at the beginning of the experiment. This was most pronounced for Trichostrongylus which represented 40 % of GIN in the sainfoin group, whereas the control group comprised only 27 % of this species. In the sainfoin fed group, the proportion of Haemonchus increased by 12 % within the 24 day experimental period whereas Teladorsagia L3 were reduced (Table 7.1). For the other species no major changes were observed within the 24 day experimental period.

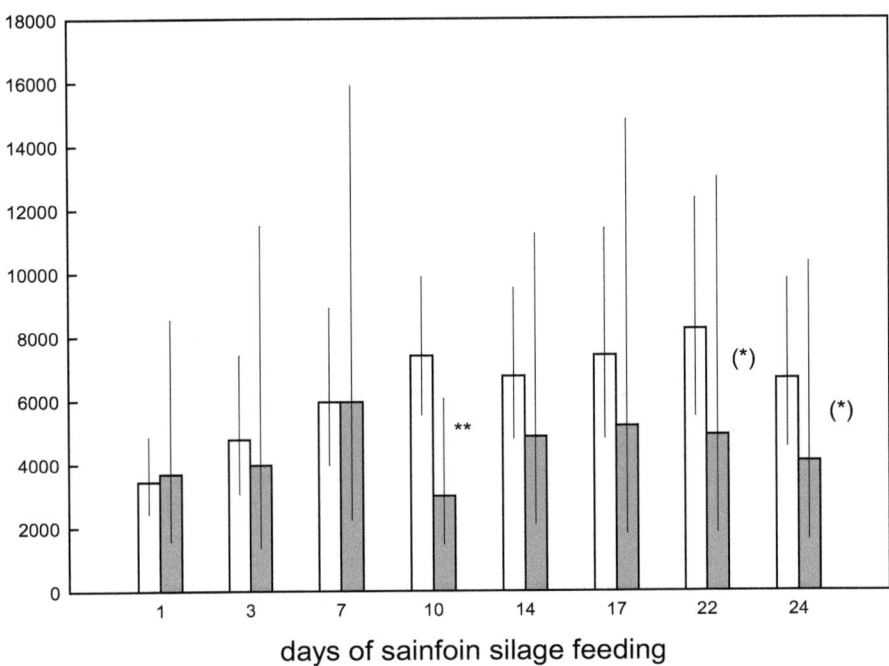

Figure 7.1. Mean faecal egg counts per gram faecal dry matter (FECDM) of control (white bars) and sainfoin (grey bars) fed animals. Error bars represent negative binomial 95 % CI. Significances: (*) P < 0.1; ** P < 0.01.

Table 7.1. Percentages of gastrointestinal nematode L3 larvae present in the cultures of the control and the sainfoin group at the start (day 1) and the end (day 24) of experimental feeding.

Day of study	Fodder	Abomasum		Small intestine	
		Haemonchus %	Teladorsagia %	Trichostrongylus %	Cooperia %
1	Control	64	4	30	2
24	Control	65	7	27	1
1	Sainfoin	42	14	41	3
24	Sainfoin	54	2	43	1

7.4. Discussion

The results of this study demonstrate that the feeding of CT-containing sainfoin silage consistently reduced the egg excretion of periparturient ewes when compared to non-CT fed control animals. The FECDM reduction was significant on Day 10 of consecutive sainfoin silage feeding and, although not significant, remained reduced until the end of the experiment. Reduced FEC in both sheep and goats carrying GIN infections and fed with sainfoin silage or hay have already been observed by others (Paolini *et al.*, 2005a, Paolini *et al.*, 2003b, Heckendorn *et al.*, 2006). In a four month controlled experiment, Paolini *et al.* (2005a) showed that the repeated distribution of sainfoin hay to adult goats carrying GIN infections comparable to the ones in the present study (e.g. mainly *Haemonchus, Teladorsagia* and *Trichostrongylus sp.*) was associated with significantly reduced FEC and an elevated packed cell volume (PCV) in the blood. In lambs carrying concurrent infections of *Haemonchus* and *Cooperia*, we found that the reduced FEC in sainfoin silage or hay fed animals was associated with reductions in adult *Haemonchus* burden when compared to controls (chapter 6).

In the present experiment, the animals in the sainfoin group were not accustomed to the novel feed. It is therefore possible that during the first days of experimental feeding the feed intake in this group was reduced when compared to the control. As a consequence, the ingested amount of CTs might have been too low to produce an anti-parasitological effect and might partly explain the delay in FECDM decrease compared to a previous experiment (chapter 6). Assuming that the significant decrease in FECDM after 10 days of consecutive sainfoin feeding was related to higher feed and CT-intake, the effect could have been expected to be sustainable until the end of the study. The variabilty in CT-content of different silage units ranging from 15 to 28 g kg^{-1} DM might partly explain the observed differences in the extent of the parasitological effect. Although it has been shown in a previous study that the fermentation process in silage production did not inactivate CTs of sainfoin, it is still possible that factors such as the density of the consolidated fodder and the air-tightness of the wrapped bales influences the stability of CT-molecules, which might explain the variability of CT-levels measured in different silage units. Alternatively, the differences in CT-content of different silage units might simply be a reflection of the fraction of sainfoin present in the unit by chance when processed from a sward containing also other non-CT containing herbs and grasses. Before sainfoin silage is to be integrated as a component of alternative control strategy against GIN, the above mentioned issues certainly deserve further attention.

In this experiment AUC based on FECDM was introduced as a quantitative measure reflecting the severity of the potential future pasture contamination when grazing with these animals. Although not significant, AUC of the sainfoin fed group was reduced by 32 % (and 65 % if calculated only for the period after onset of FECDM reduction) in the sainfoin group. Further studies need to clarify whether such a reduction actually lowers pasture contamination and represents a measurable benefit in terms of infection of naïve grazing lambs and also of re-infection of ewes. In this context, also the impact of continued grazing of ewes and lambs on sainfoin pastures following winter feeding would also be of major interest.

There has been much debate as to whether the antiparasitic effects of tanniferous plants was indirect, mediated by an improved immune response of the host against GINs or whether the parasites are directly affected by CTs (Kahn and Diaz-Hernandez, 1999, Hoste *et al.*, 2006). As in the present experiment potentially immunocompetent animals where used, the results presented here do not allow discrimination between these two hypothetical modes of CT-action. In a recent study with parasite naïve lambs, however, it has been shown that sainfoin hay or silage feeding was associated with direct antiparasitic effects (Heckendorn *et al.*, 2006). On the other hand, in non-parasitized sheep, the consumption of sainfoin has been associated with increased blood levels of essential amino acids known to be limiting in producing an effective immune response towards GINs (A. Scharenberg – personal communication). There is also evidence that the antiparasitic effect of sainfoin is not solely attributable to CTs. In an in vitro experiment, Barrau *et al.* (2005) showed that other secondary plant metabolites such as rutin and narcissin may be involved in the anthelmintic effect. Irrespective of the underlying mechanism and active compounds involved, the current results confirm some favourable antiparasitic effects associated with the consumption of sainfoin silage in naturally infected periparturient ewes. However, the potential impact of this effect on the epidemiology of GIN populations needs to be critically evaluated.

8. Overall Discussion and Conclusions

8.1. Importance of CT-dose

With respect to controlled experiments including defined GIN infections, the focus in the past has clearly been on *T. colubriformis* (small intestine) and *T. circumcincta* (abomasum). Differences between the two species in their response to tanniferous fodder plant feeding have repeatedly been observed and the question has arisen as to whether these differences where related to different sensitivities of the parasites to CTs or whether they are associated with different CT-availability in different sections of the gastrointestinal tract. In order to address this question, we included another pair of GIN in our feeding experiments: *Cooperia curticei* residing in the small intestine and *Haemonchus contortus* restricted to the abomasum. Our results suggest that a higher CT-amount is necessary to generate a measurable anthelmintic effect on the intestinal species (*Cooperia*) than on the abomasal species *Haemonchus*, (Experiments 1 & 2). The results from previous experiments with *Teladorsagia circumcincta* and/or *Trichostrongylus colubriformis* suggest that also for this pair of GIN higher CT-levels are necessary to produce an antiparasitic effect on the intestinal species than on the abomasal species (Hoste et al., 2006). Taken together, it could be speculated that the antiparasitic effects of tanniferous forages in general are achieved at lower CT-levels in the abomasum than in the small intestine and therefore would rather be organ dependent rather than GIN species related.

In both feeding experiments performed with *Haemonchus* and *Cooperia*, we found that irrespective of the tanniferous plant species administered and independent of the conservation procedure of sainfoin, the FEC specific to *Haemonchus* was significantly reduced after a 17 days feeding period. Experiment 1 indicated that this reduction was mainly associated with the observed reductions in adult *Haemonchus* burden, as no re-emergence of FEC was observed when CT-feeding was stopped. This assumption was further substantiated in the second experiment, where despite the fact that no FEC follow up was performed after the cessation of CT-feeding, no reduction in per capita fecundity was found for female *Haemonchus*. Furthermore, with respect to sainfoin, the comparison between the first and the second experiment suggested that the effect on both *Haemonchus* FEC and *Haemonchus* worm burden is stronger if higher levels of CTs are administered.

By contrast, anthelmintic effects on *Cooperia* were only observed in the second study in which increased amounts of CTs where administered compared to the first experiment (i.e. sainfoin Experiment 1: 26 g CTs kg^{-1} DM, Experiment 2: 61 g CTs kg^{-1} DM). This outcome corresponds to the previously suggested hypothesis that in general higher levels of CTs are needed to produce anthelmintic effects on nematode species of the small intestine and is further supported by studies on the intestinal species *T. colubriformis*. For example, (Athanasiadou et al., 2005) found no effects on this parasite when using plants with CT-levels of 15.8 and 15.9 g CTs kg^{-1} DM, whereas *T. colubriformis* worm burden was significantly reduced when sulla with higher CT-concentrations (~ 100 g CTs kg^{-1} CT DM) was used (Niezen et al., 1995). In order to further substantiate these hypotheses, additional parasite specific experiments including tanniferous fodder plants covering a wider range of CT-levels are needed.

Low or moderate CT-amounts (< 60 g kg^{-1} DM) have been shown to result in positive effects on both animal production aspects and anthelmintic activity in sheep and goats (Min et al., 2003a, Hoste et al., 2006). As demonstrated in Table 1 and also in the experiments described in this thesis, the anthelmintic activity was in many cases (particularly with respect to abomasal nematode species) already observed with tanniferous fodder plants containing 20-30 g CTs kg^{-1} DM. Theoretically in these cases, an additional 30-40 g CTs kg^{-1} DM could be administered to the animals without the emergence of adverse effects such as reductions in voluntary feed intake. Because CT-concentration is thought to play an important role in generating the anthelmintic effect, the addition of CTs from other sources, additional to the tanniferous plant might therefore increase the effect. Because CTs are also expressed in seed coats and pulses, it is conceivable that CTs from such sources may be included in the feed ration in the form of concentrates, and beans in particular represent a rich source of protein.

In the context of negative effects of high CT-doses, it is also worth stressing that some tanniferous plants containing very high CT-concentrations (e.g. Sericea lespedeza, 150-230 g CTs kg^{-1} DM) did not produce any adverse effects on the host (Min et al., 2003b, Lange et al., 2006), indicating that the upper limit of 60 g CTs kg^{-1} DM is variable and probably dependent on the structure of the CTs and other factors.

8.2. In vivo dose-response relation

CTs extracted from a range of tanniferous plants have repeatedly been shown to dose-dependently affect several parasitic stages of many GIN species in vitro (Molan et al., 2000c). Being aware of the inherent difficulties of showing dose dependency in vivo, the established dose-response relation between *Haemonchus* and the CT-dose revealed a tendency for a stronger anthelmintic effect on this parasite with increasing CT-content in the feed (Experiment 1). As mentioned in the introduction of this thesis, the biochemistry of CTs in the ruminant digestive tract is influenced by a multitude of factors. In particular, it is not known how CTs behave in the abomasum (where *Haemonchus* resides). However, following the theory of CT-protein complex dissociation in acidic environments, an interaction with abomasal nematodes is potentially possible. Yet, because our results are inconclusive, it might be worth progressing in this direction. If doing so, dosing trials must be performed, specifically addressing the responsiveness of individual parasite species to individual sources of CTs.

8.3. Problems related to faecal egg counts

Faecal egg counts (FEC) in parasitological studies are usually determined on a faecal fresh weight basis, assuming equal feed intake, digestibilities and faecal DM of animals. Experiments including different feeds affecting these parameters can therefore substantially influence FEC. This was clearly illustrated in all experiments performed in the course of this thesis. The direct comparison of FEC between the control and the tanniferous fodder fed groups would have always resulted in a serious overestimation of the egg counts in fresh faeces in the CT-groups because the faecal dry matter in these groups was consistently higher compared to those of control animals. Depending on the extent of differences between groups, the replacement of FEC by the measure of total daily faecal egg output or FEC expressed in relation to faecal dry matter can effectively prevent bias. Although the collection of additional data allowing the estimation of digestibilities or the measurement of faecal dry weight involve additional work and expenses, we believe that these issues are important to consider in future studies.

8.4. Further research addressing the variability of anthelmintic responses

With respect to the research dedicated to alternative strategies to control GIN parasites, tanniferous plants have received a strong impetus in the last decade. Many experiments addressing various aspects of the approach have been performed, and a substantial body of evidence has been accumulated. One important realisation and, at the same time probably the biggest challenge in this area of research, is the large variability in anthelmintic effects observed in response to tanniferous fodder consumption. This variability can be partitioned in several aspects related to the CT-composition, the plant and its cultivation, the host and the parasite. Experiments testing the anthelmintic effect of tanniferous plants must therefore ideally address all of them. An intensified collaboration between nutritionists, parasitologists, chemists and plant scientists is desirable and important in order to make progress in this area of research. In particular, the biochemical interaction of CTs at different stages of the gastrointestinal tract need to be analysed in detail in order to make progress in the identification of the exact mechanism of action of these molecules. In vitro experiments with ruminal or abomasal fluid and GIN might be a first step in approaching the important question of the mechanism of CT-action.

8.5. Inclusion of tanniferous fodder plants in rational farming systems - proposed research

As demonstrated in the previous sections of this thesis, reductions in FEC are the most coherently reported parasitological effect associated with tanniferous fodder plant feeding. Although reduced parasite egg output leads to lower pasture infectivity, the actual benefit of this in relation to tanniferous fodder plants has never been evaluated so far. Because the periparturient egg rise (PPR) in ewes in terms of pasture contamination represents a central element in the spread of the disease, we tested the potential of sainfoin silage administration during this period in naturally infected animals (Experiment 3). This study has shown that the FEC of sainfoin fed ewes was substantially reduced. As a next step, the impact of this observed FEC reduction on pasture contamination has to be evaluated. Also, experiments including the subsequent grazing of tanniferous fodder plants might provide additional parasitological benefits and might sustain the effect seen during winter indoors CT-feeding. Furthermore, such studies integrating tanniferous fodder plants in rational farming systems will provide important information with respect to the profitability and the potential practical constraints linked to the approach.

8.6. Towards an integrative approach of alternative GIN control

To date, the approach of tanniferous fodder administration cannot be expected to provide satisfactory sustainable control of GIN parasites. As outlined in the introduction of this thesis, this is also true for other alternative non-chemical GIN control strategies developed in recent years. However, because every single method has proven partial effectiveness with respect to the control of GIN parasites, the combination of strategies might be promising in terms of producing additive effects of control. Tanniferous fodder plant administration may for example be easily combined with further elements of host nutrition (i.e. protein supplementation strategies) shown to impact on resilience and resistance of the host to GIN parasite infections. Also the combination of a control strategy directed towards the free living parasitic stages (i.e. nematophagous fungi) with tanniferous fodder administration is conceivable. Until rational integrative control approaches are available, the complete abandonment of synthetic anthelmintics is not sensible. However, in view of the widespread resistance of GIN parasites against these drugs their economical and carefully targeted use is of vital importance until integrated alternative parasite control strategies are available.

9. References

Aerts, R.J., Barry, T.N. & McNabb, W.C. (1999) Polyphenols and agriculture: beneficial effects of proanthocyanidins in forages. *Agric. Ecosyst. Environ.*, **75**, 1-12.

Anthony, J.P., Fyfe, L. & Smith, H. (2005) Plant active components - a resource for antiparasitic agents? *Trends Parasitol.*, **21**, 462-468.

Arrigo, Y. (1994) *Fütterungsempfehlungen und Nährwerttabellen für Wiederkäuer*, Forschungsanstalt für Viehwirtschaftliche Produktion, Posieux, Schweiz, 3. überarbeitete Auflage, p. 328-330.

Athanasiadou, S., Kyriazakis, I., Coop, R.L. & Jackson, F. (2000a) Effects of continuous intake of condensed tannins on parasitised sheep. *Proc. Brit. Soc. Anim. Sci.*, **35**.

Athanasiadou, S., Kyriazakis, I., Jackson, F. & Coop, R.L. (2000b) Consequences of long-term feeding with condensed tannins on sheep parasitised with *Trichostrongylus colubriformis*. *Int. J. Parasitol.*, 1025-1033.

Athanasiadou, S., Kyriazakis, I., Jackson, F. & Coop, R.L. (2000c) Effects of short-term exposure to condensed tannins on adult *Trichostrongylus colubriformis*. *Vet. Rec.*, **17**, 728-732.

Athanasiadou, S., Kyriazakis, I., Jackson, F. & Coop, R.L. (2001a) Direct anthelmintic effects of condensed tannins towards different gastrointestinal nematodes of sheep: *in vitro* and *in vivo* studies. *Vet. Parasitol.*, 205-219.

Athanasiadou, S., Kyriazakis, I., Jackson, F. & Coop, R.L. (2001b) The effects of condensed tannins supplementation of foods with different protein content on parasitism, food intake and performance of sheep infected with *Trichostrongylus colubriformis*. *Brit. J. Nutr.*, **86**, 697-706.

Athanasiadou, S., Tzamaloukas, O., Kyriazakis, I., Jackson, F. & Coop, R.L. (2005) Testing for direct anthelmintic effects of bioactive forages against *Trichostrongylus colubriformis* in grazing sheep. *Vet. Parasitol.*, **127**, 233-243.

Bahaud, D., Martinez-Ortiz de Monteallo, C., Chauveau, S., Prevot, F., Torres-Acosta, J.F.J., Fouraste, I. & Hoste, H. (2006) Effects of four tanniferous plant extracts on the in vitro exsheathment of third-stage larvae of parasitic nematodes. *Parasitology*, **132**, 545-554.

Bain, R.K. (1999) Irradiated vaccines for helminth control in livestock. *Int. J. Parasitol.*, **29**, 185-191.

Baker, R.L. (1998) A review of genetic resitance to gastrointestinal nematode parasites in sheep and goats in the tropics and evidence for resistance in some sheep and goat breed in sub-humid Kenya. *Anim. Genetic Res. Inform.*, **24**, 13-30.

Barger, I.A. (1993) Influence of sex and reproductive status on susceptibility of ruminants to nematode parasitsm. *Int. J. Parasitol.*, **23**, 463-469.

Barger, I.A. (1999) The role of epidemiological knowledge and grazing management for helminth control in small ruminants. *Int. J. Parasitol.*, **29**, 41-47.

Barrau, E., Fabre, N., Fouraste, I. & Hoste, H. (2005) Effect of bioactive compounds from Sainfoin (*Onobrychis viciifolia* Scop.) on the in vitro larval migration of *Haemonchus contortus*: role of tannins and flavonol glycosides. *Parasitology*, **131**, 531-538.

Barry, T. & McNabb, W. (1999) The implications of condensed tannins on the nutritive value of temperate forages fed to ruminants. *Brit. J. Nutr.*, **81**, 263 - 272.

Beh, K.J. & Maddox, J.F. (1996) Prospects for development of genetic markers for resistance to gastrointestinal parasite infection in sheep. *Int. J. Parasitol.*, **26**, 879-897.

Bernes, G., Waller, P.J. & Christensson, D. (2000) The effect of birdsfoot trefoil (*Lotus corniculatus*) and white clover (*Trifolium repens*) in mixed pasture swards on incoming and established nematode infections in young lambs. *Acta Vet. Scand.*, **41**, 351-361.

Bila, D.M. & Dezotti, M. (2003) Pharmaceutical drugs in the environment. *Quimica Nova*, **26**, 523-530.

Borreani, G., Peiretti, P.G. & Tabacco, E. (2003) Evolution of yield and quality of sainfoin (*Onobrychis viciifolia* Scop.) in the spring growth cycle. *Agronomie*, **23**, 193-201.

Brunsdon, R.V. & Vlassoff, A. (1982) Production and parasitological responses of lambs exposed to differing low levels of trichostrongylid larvae on pasture. *NZ J. Exp. Agric.*, **10**, 391-394.

Coop, R.L., Graham, R.B., Jackson, F., Wright, S.E. & Angus, K.W. (1985) Effect of experimental *Ostertagia circumcincta* infection on the performance of grazing lambs. *Res. Vet. Sci.,* **38,** 282-287.

Coop, R.L. & Holmes, P.H. (1996) Nutrition and parasite interaction. *Int. J. Parasitol.,* **26,** 951-962.

Coop, R.L. & Kyriazakis, I. (1999) Nutrition-parasite interaction. *Vet. Parasitol.,* **84,** 187-204.

Coop, R.L. & Kyriazakis, I. (2001) Influence of host nutrition on the development and consequences of nematode parasitism in ruminants. *Trends Parasitol.,* **17,** 325-330.

Eysker, M., Bakker, N., van der Hall, Y.A., van Hecke, I., Kooyman, F.N.J., van der Linden, D., Schrama, C. & Ploeger, H.W. (2006) The impact of daily *Duddingtonia flagrans* application to lactating ewes on gastrointestinal nematodes infections in their lambs in the Netherlands. *Vet. Parasitol.,* **141,** 91-100.

Faedo, M., Larsen, M. & Waller, P.J. (1997) The potential of nematophagous fungi to control the free-living stages of nematode parasites of sheep: Comparison between Australian isolates of *Arthrobotrys spp.* and *Duddingtonia flagrans. Vet. Parasitol.,* **72,** 149-155.

Geary, T.G., Sangster, N.C. & Thompson, D.P. (1999) Frontiers in anthelmintic pharmacology. *Vet. Parasitol.,* **84,** 275-295.

Githiori, J.B., Athanasiadou, S. & Thamsborg, S. (2006) Use of plants in novel approaches for control of gastrointestinal helminths in livestock whith emphasis on small ruminants. *Vet. Parasitol.,* **139,** 308-320.

Githiori, J.B., Hoglund, J., Waller, P.J. & Baker, R.L. (2003) Evaluation of anthelmintic properties of extracts from some plants used as livestock dewormers by pastoralist and smallholder farmers in Kenya against *Heligmosomoides polygyrus* infections in mice. *Vet. Parasitol.,* **118,** 215-226.

Githiori, J.B., Hoglund, J., Waller, P.J. & Baker, R.L. (2004) Evaluation of anthelmintic properties of some plants used as livestock dewormers against *Haemonchus contortus* infections in sheep. *Parasitology,* **129,** 245-253.

Gray, G.D. & Gill, H.S. (1993) Host genes, parasites and parasitic infections. *Int. J. Parasitol.,* **23,** 485-494.

Grundhofer, P., Niemetz, R., Schilling, G. & Cross, G.G. (2001) Biosynthesis and subcellular distribution of hydrolysable tannins. *Phythochemistry,* **57,** 915-927.

Häring, D.A., Suter, D., Amrhein, N. & Lüscher, A. (2007) Biomass allocation is an important determinant of the tannin concentration in growing plants. *Ann. Bot.,* 99, *111-120.*

Haslam, E. (1996) Natural polyphenols (vegetable tannins) as drugs: possible modes of action. *J. Nat. Prod.,* **59,** 205-215.

Heckendorn, F., Häring, D.A., Maurer, V., Zinsstag, J., Langhans, W. & Hertzberg, H. (2006) Effect of sainfoin (*Onobrychis viciifolia*) silage and hay on established populations of *Haemonchus contortus* and *Cooperia curticei* in lambs. *Vet. Parasitol.,* **142,** 293-300.

Hördegen, P., Hertzberg, H., Langhans, W. & Maurer, V. (2003) The anthelmintic efficacy of five plant products against gastrointestinal trichostrongylids in artificially infected lambs. *Vet. Parasitol.,* **117,** 51-60.

Hoskin, S.O., Wilson, P.R., Barry, T.N., Charleston, W.A.G. & Waghorn, G.C. (2000) Effect of forage legumes containing condensed tannins on lungworm (*Dictyocaulus sp.*) and gastrointestinal parasitism in young red deer (*Cervus elaphus*). *Res. Vet. Sci.,* **68,** 223-230.

Hoste, H., Gaillard, L. & Le Frileux, Y. (2005) Consequences of the regular distribution of sainfoin hay on gastrointestinal parasitism with nematodes and milk production in dairy goats. *Small Ruminant Res.,* **59,** 265-271.

Hoste, H., Jackson, F., Athanasiadou, S., Thamsborg, S. & Hoskin, S.O. (2006) The effect of tannin-rich plants on parasitic nematodes in ruminants. *Trends Parasitol.,* **22,** 253-261.

Jackson, F. & Coop, R.L. (2000) The development of anthelmintic resistance in sheep nematodes. *Parasitology,* **120,** 95-107.

Jones, W.T. & Mangan, J.L. (1977) Complexes of the Condensed Tannins of Sainfoin *Onobrychis viciifolia* with fraction 1 leaf protein and with submaxillary muco protein and their reversal by poly ethylene glycol and pH. *J. Sci. Food Agric.,* **28,** 126-136.

Kabasa, J.D., Opuda-Asibo, J. & ter Meulen, U. (2000) The effect of oral administration of polyethylene glycol on faecal helminth egg counts in pregnant goats grazed on browse containing condensed tannins. *Trop. Anim. Health Prod.,* **32,** 73-86.

Kahn, L. & Diaz-Hernandez, A. (1999) Tannins with anthelmintic properties. *ACIAR Proceedings,* No **92,** 130 - 139.

Knox, D.P. (2000) Development of vaccines against gastrointestinal nematodes. *Parasitology,* **120,** S43-S61.

Lange, K.C., Olcott, D.D., Miller, J.E., Mosjidis, J.A., Terrill, T.H., Burke, J.M. & Kearney, M.T. (2006) Effect of sericea lespedeza (*Lespedeza cuneata*) fed as hay, on natural and experimental *Haemonchus contortus* infections in lambs. *Vet. Parasitol.,* **141,** 273-278.

Larsen, M. (1999) Biological control of helminths. *Int. J. Parasitol.,* **29**, 139-146.

Larsen, M., Faedo, M., Waller, P.J. & Hennessy, D.R. (1998) The potential of nematophagous fungi to control the free-living stages of nematode parasites of sheep: Studies with *Duddingtonia flagrans. Vet. Parasitol.,* **76**, 121-128.

Larsen, M., Nansen, P., Wolstrup, J., Gronvold, J., Henriksen, S.A. & Zorn, A. (1995) Biological control of trichostrongyles in calves by the fungus *Duddingtonia flagrans* fed to animals under natural grazing conditions. *Vet. Parasitol.,* **60**, 321-330.

Le Jambre, L.F., Dobson, R.J., Lenane, I.J. & Barnes, E.H. (1999) Selection for anthelmintic resistance by macrocyclic lactones in *Haemonchus contortus. Int. J. Parasitol.,* **29**, 1101-1111.

Lowry, J.B., McSweeney, C.S. & Palmer, B. (1996) Changing perceptions of the effect of plant phenolics on nutrient supply in the ruminant. *Aust. J. Agr. Res.,* **47**, 829-842.

Luck, G., Liao, H., Murray, N.J., Grimmer, H.R., Warminski, E.E., Williamson, M.P., Lilley, T.H. & Haslam, E. (1994) Polyphenols, astringency and proline-rich proteins. *Phytochemistry,* **37**, 357-371.

Lüscher, A., Häring, D.A., Heckendorn, F., Scharenberg, A., Dohme, F., Maurer, V. & Hertzberg, H. (2005) Use of tanniferous plants against gasto-intestinal nematodes in ruminants. *In: Researching sustainable systems. Proceedings of the 15th IFOAM Organic World Congress, 21-23.9.2005, Adelaide, South Australia,* 272-276.

MAFF (1986) Manual of Veterinary Parasitological Techniques. *Technical Bulletin No. 18, HMSO, London,* 36-40.

Marley, C.L., Cook, R., Barrett, J., Keatinge, R., Lampkin, N.H. & McBride, S.D. (2003a) The effect of dietary forage on the development and survival of helminth parasites in ovine faeces. *Vet. Parasitol.,* **118**, 93-107.

Marley, C.L., Cook, R., Keatinge, R., Barrett, J. & Lampkin, N.H. (2003b) The effect of birdsfoot trefoil (*Lotus corniculatus*) and chicory (*Cichorium intybus*) on parasite intensities and performance of lambs naturally infected with helminth parasites. *Vet. Parasitol.,* **112**, 147-155.

Mayes, R.W., Lamb, C.S. & Colgrove, P.M. (1986) The use of dosed and herbage n-alkanes as markers for the determination of herbage intake. *J. Agric. Sci.,* **107**, 161-170.

McNabb, W.C., Waghorn, G.C., Peters, J.S. & Barry, T.N. (1996) The effect of condensed tannins in Lotus pedunculatus on the solubilization and degradation of ribulose-1,5-bisphosphate carboxylase (EC 4.1.1.39; Rubisco) protein in the rumen and the sites of Rubisco digestion. *Brit. J. Nutr.,* **76**, 535-549.

Michel, J.F. (1974) Arrested development of nematodes and some related phenomena. *Adv. Parasit.,* **12**, 279-309.

Michel, J.F. (1976) The epidemiology and control of some nematode infections in grazing animals. *Adv. Parasit.,* **14**, 355-397.

Min, B.R., Barry, T.N., Attwood, G.T. & McNabb, W.C. (2003a) The effect of condensed tannins on the nutrition and health of ruminants fed fresh temperate forages: A review. *Anim. Feed Sci. Tech.,* **106**, 3-19.

Min, B.R. & Hart, S.P. (2002) Tannins for suppression of internal parasites. *J. Anim. Sci.,* **85**, E102-E109.

Min, B.R., Hart, S.P., Miller, D., Tomita, G.M., Loetz, E. & Sahlu, T. (2005) The effect of grazing forage containing condensed tannins on gastro-intestinal parasite infection and milk composition in Angora does. *Vet. Parasitol.,* **130**, 105-113.

Min, B.R., Miller, D., Hart, S.P., Tomita, G., Loetz, E. & Sahlu, T. (2003b) Direct effects of condensed tannins on gastrointestinal nematodes in grazing Angora goats. *J. Anim. Sci.,* **81**, 23-24.

Min, B.R., Pomroy, W.E., Hart, S.P. & Sahlu, T. (2004) The effect of short-term consumption of a forage containing condensed tannins on gastro-intestinal nematode parasite infections in grazing wether goats. *Small Ruminant Res.,* **51**, 279-283.

Molan, A.L., Alexander, R., Brookes, I.M. & McNabb, W.C. (2000a) Effects of an extract of sulla (*Hedysarium coronarium*) containing condensed tannins on the migration of three sheep gastrointestinal nematodes in vitro. *Proc N. Z. Soc. Anim. Prod.,* **60**, 21-25.

Molan, A.L., Attwood, G.T., Min, B.R. & McNabb, W.C. (2001) The effect of condensed tannins from *Lotus pedunculatus* and *Lotus corniculatus* on the growth of proteolytic rumen bacteria in vitro and their possible mode of action. *Can. J. Microbiol.,* **47**, 626-633.

Molan, A.L., Duncan, A., Barry, T.N. & McNabb, W.C. (2000b) Effects of condensed tannins and sesquiterpene lactones extracted from chicory on the viability of deer lungworm larvae. *Proc. NZ Soc. Anim. Prod.,* **60**, 26-29.

Molan, A.L., Duncan, A.J., Barry, T.N. & McNabb, W.C. (2003a) Effects of condensed tannins and crude sesquiterpene lactones extracted from chicory on the motility of larvae of deer lungworm and gastrointestinal nematodes. *Parasitol. Int.,* **52**, 209-218.

Molan, A.L., Meagher, L.P., Spencer, P.A. & Sivakumaran, S. (2003b) Effect of flavan-3-ols on in vitro egg hatching, larval development and viability of infective larvae of *Trichostrongylus colubriformis*. *Int. J. Parasitol.*, **33**, 1691-1698.

Molan, A.L., Waghorn, G.C., Min, B.R. & McNabb, W.C. (2000c) The effect of condensed tannins from seven herbages on *Tricostrongylos colubriformis* larval migration *in vitro*. *Folia Parasit.*, **47**, 39-44.

Mosjidis, C.O., Peterson, C.M. & Mosjidis, J.A. (1990) Developmental Differences in the Location of Polyphenols and Condensed Tannins in Leaves and Stems of Sericea Lespedeza (*Lespedeza cuneata*). *Ann. Bot.*, **65**, 355-360.

Moss, R.A. & Vlassoff, A. (1993) Effect of herbage species on gastro-intestinal roundworm populations and their distribution. *NZ J. Agric. Res.*, **36**, 371-375.

Mueller Harvey, I. (2001) Analysis of hydrolysable tannins. *Anim. Feed Sci. Tech.*, **91**, 3-20.

Mueller Harvey, I. & McAllan, A.B. (1992) Tannins, their biochemistry and nutritional properties. In: *Advances in Plant Cell Biochemistry and Biotechnology, Vol. 1*, edited by: *I. M. Morrison*, pp. 151-217.

Niezen, J.H., Charleston, W.A.G., Robertson, H.A., Shelton, D., Waghorn, G.C. & Green, R. (2002a) The effect of feeding sulla (*Hedysarum coronarium*) or lucerne (*Medicago sativa*) on lamb parasite burdens and development of immunity to gastrointestinal nematodes. *Vet. Parasitol.*, **105**, 229-245.

Niezen, J.H., Robertson, H.A., Waghorn, G.C. & Charleston, W.A. (1998a) Production, faecal egg counts and worm burdens of ewe lambs which grazed six contrasting forages. *Vet. Parasitol.*, **80**, 15-27.

Niezen, J.H., Waghorn, G.C. & Charleston, W.A.G. (1998b) Establishment and fecundity of *Ostertagia circumcincta* and *Trichostrongylus colubriformis* in lambs fed lotus (*Lotus pedunculatus*) or perennial ryegrass (*Lolium perenne*). *Vet. Parasitol.*, **78**, 13-21.

Niezen, J.H., Waghorn, G.C., Graham, T., Carter, J.L. & Leathwick, D.M. (2002b) The effect of diet fed to lambs on subsequent development of *Trichostrongylus colubriformis* larvae in vitro and on pasture. *Vet. Parasitol.*, **105**, 269-283.

Niezen, J.H., Waghorn, T.S., Charleston, W.A. & Waghorn, G.C. (1995) Growth and gastrointestinal nematode parasitism in lambs grazing lucerne (*Medicago sativa*) or sulla (*Hedysarum coronarium*) which contains condensed tannins. *J. Agric. Sci.*, **125**, 281-289.

Niezen, J.H., Waghorn, T.S., Raufaut, K., Robertson, H.A. & McFarlane, R.G. (1994) Lamb weight gain and faecal egg count when grazing one of seven herbages and dosed with larvae for six weeks. *Proc. NZ Soc. Anim. Prod.*, **54**, 15-18.

Paolini, V., Bergeaud, J.P., Grisez, C., Prevot, F., Dorchies, P. & Hoste, H. (2003a) Effects of condensed tannins on goats experimentally infected with *Haemonchus contortus*. *Vet. Parasitol.*, **113**, 253-261.

Paolini, V., De La Farge, F., Prevot, F., Dorchies, P. & Hoste, H. (2005a) Effects of the repeated distribution of sainfoin hay on the resistance and the resilience of goats naturally infected with gastrointestinal nematodes. *Vet. Parasitol.*, **127**, 277-283.

Paolini, V., Dorchies, P. & Hoste, H. (2003b) Effects of sainfoin hay on gastrointestinal nematode infections in goats. *Vet. Rec.*, **152**, 600-601.

Paolini, V., Fouraste, I. & Hoste, H. (2004) In vitro effects of three woody plant and sainfoin extracts on 3rd-stage larvae and adult worms of three gastrointestinal nematodes. *Parasitology*, **129**, 69-77.

Paolini, V., Frayssines, A., De La Farge, F., Dorchies, P. & Hoste, H. (2003c) Effects of condensed tannins on established populations and on incoming larvae of *Trichostrongylus colubriformis* and *Teladorsagia circumcincta* in goats. *Vet. Res.*, **34**, 331-339.

Paolini, V., Prevot, F., Dorchies, P. & Hoste, H. (2005b) Lack of effects of quebracho and sainfoin hay on incoming third-stage larvae of *Haemonchus contortus* in goats. *Vet. J.*, **170**, 260-263.

Parkins, J.J. & Holmes, P.J. (1989) Effects of gastrointestinal helminth parasites on ruminant nutrition. *Nutr. Res. Rev.*, **2**, 227-246.

Perry, B.D. & Randolph, T.F. (1999) Improving the assessment of the economic impact of parasitic diseases and of their control in production animals. *Vet. Parasitol.*, **84**, 145-168.

Pomroy, W.E. & Adlington, B.A. (2006) Efficacy of short-term consumption of sulla (*Hedysarium coronarium*) to young goats against a mixed burden of gastrointestinal nematodes. *Vet. Parasitol.*, **136**, 363-366.

Poppi, D.P., Macrae, J.C., Brewer, A. & Dewey, P.J.S. (1985) Calcium and phosphorus absorption in lambs exposed to *Trichostrongylus colubriformis*. *J. Comp. Pathol.*, **95**, 453-464.

Reed, J.D. (1995) Nutritional toxicology of tannins and related polyphenols in forage legumes. *J. Anim. Sci.*, **73**, 1516-1528.

Rehbein, S., Kollmannsberger, M., Visser, M. & Winter, R. (1996) The helminth fauna of slaughtered sheep from upper Bavaria: 1. Species composition, prevalence and wormcounts. *Berl. Münch. Tierärztl. Wschr.,* **109**, 161-167.

Rehbein, S., Visser, M. & Winter, H. (1998) Endoparasitic infections in sheep from the Swabian alb. *Dtsch. Tierärztl. Wschrift,* **105**, 419-424.

Scales, G.H., Knight, T.L. & Saville, D.J. (1994) Effect of herbage species and feeding level on internal parasites and production performance of grazing lambs. *NZ J. Agric. Res.,* **38**, 237-247.

Schillhorn Van Veen, T.W. (1997) Sense or nonsense? Traditional methods of animal parasitic disease control. *Vet. Parasitol.,* **71**, 177-194.

Schmidt, U. (1971) Parasitologische Kotuntersuchung durch ein neues Verdünnungsverfahren. *Tieraerztl. Umsch.,* 229-230.

Shaik, S.A., Terrill, T.H., Miller, J.E., Kouakou, B., Kannan, G., Kaplan, R.M., Burke, J.M. & Mosjidis, J.A. (2006) Sericea lespedeza hay as a natural deworming agent against gastrointestinal nematode infections in goats. *Vet. Parasitol.,* **139**, 150-157.

Silanikove, N., Gilboa, N., Perevolotsky, A. & Nitsan, Z. (1996) Goats fed tannin-containing leaves do not exhibit toxic syndromes. *Small Ruminant Res.,* **21**, 195-201.

Smith, W.D. (1999) Prospects for vaccines of helminth parasites of grazing ruminants. *Int. J. Parasitol.,* **29**, 17-24.

Sonstegard, T.S. & Gasbarre, L.C. (2001) Genomic tools to improve parasite resistance. *Vet. Parasitol.,* **101**, 387-403.

Stear, M.J. & Murray, M. (1994) Genetic resistance to parasitic disease: Particularly of resistance in ruminants to gastrointestinal nematodes. *Vet. Parasitol.,* **54**, 161-176.

Stepek, G., Behnke, J.M., Buttle, D.J. & Duce, I.R. (2004) Natural plant cysteine proteinases as anthelmintics? *Trends Parasitol.,* **20**, 322-327.

Sykes, A.R. (1994) Parasitism and production in farm animals. *Anim. Prod.,* **59**, 155-172.

Symons, L.E.A. & Jones, W.O. (1975) Skeletal muscle, liver and wool protein synthesis by sheep infected by the nematode *Trichostrongylus colubriformis. Aust. J. Agric. Res.,* **26**, 1063-1072.

Terrill, T.H., Rowan, A.M., Douglas, G.B. & Barry, T.N. (1992) Determination of extractable and bound condensed tannin concentrations in forage plants, protein concentrate meals and cereal grains. *J. Sci. Food Agric.,* 321-329.

Terrill, T.H., Waghorn, G.C., Woolley, D.J., McNabb, W.C. & Barry, T.N. (1994) Assay and digestion of 14C- labelled condensed tannins in the gastrointestinal tract of sheep. *Brit. J. Nutr.,* **72**, 467-477.

Thamsborg, S.M., Mejer, H., Bandier, M. & Larsen, M. (2004) Influence of different forages on gastrointestinal nematode infections in grazing lambs. *In: Proceedings of the 19th International Conference of the World Association for the Advancement of Veterinary Parasitology, 10-14 August, New Orleans,* p. 189.

Thamsborg, S.M., Roepstorff, A. & Larsen, M. (1999) Integrated and biological control of parasites in organic and conventional production systems. *Vet. Parasitol.,* **84**, 169-186.

Tilley, J.M.A. & Terry, R.A. (1963) A two stage in vitro digestion of forage crops. *J. Brit. Grassl. Assoc.,* **18**, 104-111.

Torgerson, P.R., Schnyder, M. & Hertzberg, H. (2005) Detection of anthelmintic resistance: a comparison of mathematical techniques. *Vet. Parasitol.,* **128**, 291-298.

Tzamaloukas, O., Athanasiadou, S., Kyriazakis, I., Jackson, F. & Coop, R.L. (2005) The consequences of short-term grazing of bioactive forages on established adult and incoming larvae populations of *Teladorsagia circumcincta* in lambs. *Int. J. Parasitol.,* **35**, 329-335.

Urquhart, G.M., Jarrett, W.F.H., Jennings, F.W., McIntyre, W.I.M. & Mulligan, W. (1966a) Immunity to *Haemonchus contortus* Infection: Relationship between age and successful Vaccination with irradiated larvae. *Am. J. Vet. Res.,* **27**, 1645-1648.

Urquhart, G.M., Jarrett, W.F.H., Jennings, F.W., Molntyre, W.I.M., Mulligan, W. & Sharp, N.C.C. (1966b) Immunity to *Haemonchus contortus* Infection: Failure of X-irradiated larvae to immunise young Lambs. *Am. J. Vet. Res.,* **27**, 1641-1643.

Van Houtert, M.F.J. & Sykes, A.R. (1996) Implications of nutrition for the ability of ruminants to withstand gastrointestinal nematode infections. *Int. J. Parasitol.,* **26**, 1151-1167.

Waghorn, G.C. & McNabb, W.C. (2003) Consequences of plant phenolic compounds for productivity and health of ruminants. *Proc. Nutr. Soc.,* **62,** 383-392.

Waller, P.J. (1999) International approaches to the concept of integrated control of nematode parasites of livestock. *Int. J. Parasitol.,* **29,** 155-164.

Waller, P.J., Bernes, G., Thamsborg, S., Sukura, A., Richter, S.H., Ingebrigtsen, K. & Hoglund, J. (2001) Plants as de-worming agents of livestock in the Nordic countries: historical perspective, popular beliefs and prospects for the future. *Acta Vet. Scand.,* **42,** 31-44.

Waller, P.J. & Thamsborg, S. (2004) Nematode control in 'green' ruminant production systems. *Trends Parasitol.,* **20,** 493-497.

Waterman, P.G. (1999) The tannins - an overview. *In: Tannins in Livestock and Human Nutrition (ed. Brooker J.D.). Proceedings of International Workshop, Adelaide, pp. 10-13, Australian Centre for International Agricultural research.*

Williams, J.C. (1997) Anthelmintic treatment strategies: Current status and future. *Vet. Parasitol.,* **72,** 461-470.

Wilson, K., Grenfeld, B.T. & Shaw, D.J. (1996) Analysis of aggregated parasite distributions. *Funct. Ecol.,* **10,** 592-601.

Zhu, J., Filippich, L.J. & Alsalami, M.T. (1992) Tannic acid intoxication in sheep and mice. *Res. Vet. Sci.,* **53,** 280-292.

Publications - Felix Heckendorn

Peer reviewed publications (in addition to the publications included in this thesis)

Heckendorn, F., N'Goran E.K., Felger, I., Vounatsou, P., Yapi, A., Oettli, A., Marti, H.P., Dobler, M., Traore, M., Lohourignon, K.L. & Lengeler, C. (2002) Species-specific field testing of *Entamoeba spp*. In an area of high endemicity. Trans. R. Soc. Trop. Med. Hyg: 96, 521-528.

Conference abstracts

Heckendorn, F., Maurer, V., Langhans, W. & Hertzberg, H. (2006). Esparsette (*Onobrychis viciifolia*) als mögliche bioaktive Futterpflanze für die Kontrolle von Magen-Darm-Strongyliden bei Schafen. 22. Jahrestagung der DGP, 22.-25.2.2006, VMU Wien, S.55

Heckendorn, F., Maurer, V., Senn, M. & Hertzberg, H. (2006) Esparsette (*Onobrychis viciifolia*) als mögliche Futterpflanze zur Kontrolle von Magen-Darm-Strongyliden bei Schafen. Tagung der Deutschen Veterinärmedizinischen Gesellschaft (DVG), 7.-9.5.2006, Wetzlar, 50.

Häring, D.A., **Heckendorn, F.**, Scharenberg, A., Amrhein, N., Dohme, F., Kreuzer, M., Langhans, W., Lüscher, A., Maurer, V., Suter, D. & H. Hertzberg (2005). Tanniniferous plants as a control agent against gastro-intestinal nematodes in ruminants. *In: Quality legume-based forage systems for contrasting environments. Proceedings of the COST 852 workshop, 10 – 12 November 2005, Grado, Italy.*

Maurer, V. & **Heckendorn, F.** (2006). Le marché vétérinaire des plantes médicinales: Contrôle des vers gastro-intestinaux. Colloque Valplantes, Sion, 7-8 Septembre 2006.

Lüscher, A, Häring, D.A, **Heckendorn, F.**, Scharenberg, A., Dohme, F., Maurer, V. & H. Hertzberg (2005). Use of tanniferous plants against gastro-intestinal nematodes in ruminants. *In:* Researching sustainable systems. Proceedings of the 15th IFOAM Organic World Congress, 21.-23.9.2005, Adelaide, South Australia, 272-276.

Hertzberg, H., Noto, F., Figi, R. & **Heckendorn, F.** (2004) Control of gastrointestinal nematodes in organic beef cattle through grazing management. Proceedings of the 2nd SAFO workshop, Witzenhausen, Germany

Scott-Baird, E., **Heckendorn, F.**, Maurer, V., Jackson, F., Leifert, C., Edwards, S. & Butler, G. (2006) Evaluation of six plant extracts for potential anthelmintic properties under in vitro conditions – Proceedings of the 'Ethnoveterinary Conference - Harvesting Knowledge, Pharming Opportunities', Writtle College, UK.

Other Publications

Heckendorn, F., Maurer, V. & Hertzberg, H. (2006) Kontrolle von Magen-Darm-Strongyliden bei Schafen. Vet-MedReport 30, Sonderausgabe V4: 4-5.

Heckendorn, F. (2005). Kondensierte Tannine – Eine Möglichkeit zur Kontrolle von Magen-Darm-Würmern? Forum 1 / 2, 11-16.

Hertzberg, H., Maurer, V., **Heckendorn, F.**, Wanner, A., Gutzwiller, A. & Mosimann, E. (2007) Wurm-Befall bei Jungrindern unter trockenen Weidebedingungen. Agrarforsch. 14, 28-33.

Südwestdeutscher Verlag für Hochschulschriften

Wissenschaftlicher Buchverlag bietet
kostenfreie
Publikation
von
Dissertationen und Habilitationen

Sie verfügen über eine wissenschaftliche Abschlußarbeit zu aktuellen oder zeitlosen Fragestellungen, die hohen inhaltlichen und formalen Anspruchen genügt, und haben **Interesse an einer honorarvergüteten Publikation?**

Dann senden Sie bitte erste Informationen über Ihre Arbeit per Email an: info@svh-verlag.de.

Unser Außenlektorat meldet sich umgehend bei Ihnen.

Südwestdeutscher Verlag für Hochschulschriften
Aktiengesellschaft & Co. KG
Dudweiler Landstr. 99
D – 66123 Saarbrücken
www.svh-verlag.de

Printed by Books on Demand GmbH, Norderstedt / Germany